Lecture Notes in Computer Science   12930

More information about this subseries at http://www.springer.com/series/8637

Abdelkader Hameurlain ·
A Min Tjoa (Eds.)

Transactions on Large-Scale Data- and Knowledge-Centered Systems L

 Springer

Editors-in-Chief
Abdelkader Hameurlain
IRIT, Paul Sabatier University
Toulouse, France

A Min Tjoa ⓘ
IFS, Technical University of Vienna
Vienna, Austria

ISSN 0302-9743 ISSN 1611-3349 (electronic)
Lecture Notes in Computer Science
ISSN 1869-1994 ISSN 2510-4942 (electronic)
Transactions on Large-Scale Data- and Knowledge-Centered Systems
ISBN 978-3-662-64552-9 ISBN 978-3-662-64553-6 (eBook)
https://doi.org/10.1007/978-3-662-64553-6

This Springer imprint is published by the registered company Springer-Verlag GmbH, DE
part of Springer Nature
The registered company address is: Heidelberger Platz 3, 14197 Berlin, Germany

Preface

This volume contains five fully revised selected regular papers, covering a wide range of very hot topics in the fields of data anonymization, quasi-identifier discovery methods, symbolic time series representation, detection of anomalies in time series, data quality management in biobanks, and multi-agent technology in the design of intelligent systems for maritime transport.

We would like to sincerely thank the editorial board and the external reviewers for thoroughly refereeing the submitted papers and ensuring the high quality of this volume.

Special thanks go to Shaoyi Yin for her valuable help in mastering the new system used for the submission and evaluation of papers.

September 2021

Abdelkader Hameurlain
A Min Tjoa

Organization

Editors-in-Chief

Abdelkader Hameurlain	Paul Sabatier University, IRIT, France
A Min Tjoa	Technical University of Vienna, IFS, Austria

Editorial Board

Reza Akbarinia	Inria, France
Dagmar Auer	FAW, Austria
Djamal Benslimane	Claude Bernard University Lyon 1, France
Stéphane Bressan	National University of Singapore, Singapore
Mirel Cosulschi	University of Craiova, Romania
Dirk Draheim	Tallinn University of Technology, Estonia
Johann Eder	Alpen-Adria University of Klagenfurt, Austria
Anna Formica	National Research Council, Italy
Shahram Ghandeharizadeh	University of Southern California, USA
Anastasios Gounaris	Aristotle University of Thessaloniki, Greece
Theo Härder	Technical University of Kaiserslautern, Germany
Sergio Ilarri	University of Zaragoza, Spain
Petar Jovanovic	BarcelonaTech, Spain
Aida Kamišalić Latifić	University of Maribor, Slovenia
Dieter Kranzlmüller	Ludwig-Maximilians-Universität München, Germany
Philippe Lamarre	INSA Lyon, France
Lenka Lhotská	Technical University of Prague, Czech Republic
Vladimir Marik	Technical University of Prague, Czech Republic
Jorge Martinez Gil	Software Competence Center Hagenberg, Austria
Franck Morvan	Paul Sabatier University, IRIT, France
Torben Bach Pedersen	Aalborg University, Denmark
Günther Pernul	University of Regensburg, Germany
Soror Sahri	LIPADE, Paris Descartes University, France
Shaoyi Yin	Paul Sabatier University, IRIT, France
Feng "George" Yu	Youngstown State University, USA

External Reviewers

Mahmoud Barhamgi	Claude Bernard University Lyon 1, France
Frédéric Migeon	Paul Sabatier University, IRIT, France
Riad Mokadem	Paul Sabatier University, IRIT, France
Joseph Vella	University of Malta, Malta

Contents

A Parallel Quasi-identifier Discovery Scheme for Dependable Data Anonymisation

Nikolai J. Podlesny$^{(\boxtimes)}$, Anne V. D. M. Kayem$^{(\boxtimes)}$, and Christoph Meinel

Hasso Plattner Institute at the University of Potsdam, Potsdam, Germany
Nikolai.Podlesny@student.hpi.de, {Anne.Kayem,Christoph.Meinel}@hpi.de

Abstract. Quasi-identifiers (QIDs) are attribute combinations that can be used to discover hidden personal identifying information from an anonymised dataset. Typically, the information drawn from such QIDs can then be combined with more publicly accessible datasets to discover sensitive information (e.g. medical conditions, financial status, criminal history, ...). Research on data anonymisation has therefore proposed various algorithms to discover and transform quasi-identifiers efficiently to prevent re-identification. However, all existing algorithms are inefficient and fail to prevent re-identification attacks on large real-world high dimensional datasets successfully. This paper presents a quasi-identifier discovery algorithm that combines parallelism with an efficient search technique to find all minimal quasi-identifiers in a given dataset. As a further step, we present an adversary model based on the enumeration problem of discovering unique column combinations in a dataset. We demonstrate that our quasi-identifier discovery algorithm is secure to re-identification attacks based on this adversarial model, even in the presence of large high-dimensional datasets that change dynamically. Our empirical results show that our algorithm not only scales well to large high-dimensional datasets but exploits its parallelisability on GPU (Graphics Processing Unit) architectures to prevent re-identification even in the presence of a powerful adversary equipped with similar high-performance computing processing power. Furthermore, our results show that the proposed GPU algorithm offers up to 100x times speedup over the algorithm's CPU version.

1 Introduction

A quasi-identifier (QID) described as $Q \rightarrow_{ID} r_i$ is a combination of attributes that contain values that can be used to uniquely identify a record in dataset [34]. So in essence, $Q \rightarrow_{ID} r_i$ indicates that QID (Q) identifies record r_i uniquely. Identifying quasi-identifiers efficiently is crucial to generating privacy preserving datasets. More importantly, being able to do so even in the presence of auxiliary data and in real-time, can make an important contribution to the problem of anonymising highly sensitive datasets (e.g. health, financial, and criminal data).

Other applications for efficient quasi-identifier discovery arise in data processing and analytics operations such as, inter-application data sharing [10]

© Springer-Verlag GmbH Germany, part of Springer Nature 2021
A. Hameurlain and A. M. Tjoa (Eds.): TLDKS L, LNCS 12930, pp. 1–24, 2021.
https://doi.org/10.1007/978-3-662-64553-6_1

(e.g. social media platforms and third-party applications), data integration [3], and fraud detection [52]. While these use-cases exist, to the best of our knowledge, efficient QID discovery is never explicitly mentioned as the crux of the issue of anonymising highly sensitive datasets. One possible reason, is that privacy-preserving techniques tend to focus primarily on detecting and eliminating Personal Identifying Information (PII). For example, a person's *firstname* might determine his/her *gender* or *birthdate* might determine *age*. However, QIDs add another dimension to this problem in that single attributes might be benign but when combined with other attributes become exploitable for re-identification attacks. For example, removing *firstname* from a dataset reduces but does not remove the risk of re-identifying an individual because it might still be possible to re-identify the person successfully by using information drawn from a combination of *gender, age, race* and *postcode*. In this case $\{gender, age, race, postcode\} \rightarrow_{ID} Person$. As such, QID discovery algorithms are essential to revealing all the QIDs in a given dataset.

Problem Statement. The importance of the QID discovery problem has led to the proposal of a plethora of solutions centred around finding ways of circumventing attacks due to inferences drawn from attribute value combinations [20, 34, 35, 41, 42, 51]. However, all of the existing solutions face three fundamental problems.

1. **QID Discovery on Large High-Dimensional Data.** None of the existing algorithms is able to process large high- dimensional real-world datasets (e.g. datasets with more than 100 attributes and a million records) efficiently. This is because the cost of transforming the data (e.g. using generalisation, suppression, and/or perturbation) grows with the size of the dataset. A further reason is that the transformation mechanisms optimise for either equivalence class sizes or distributions of sensitive values and so, certain QIDs can get overlooked resulting in re-identification [22, 25, 48].
   In essence, the QID discovery problem is NP-hard. More specifically, as Podlesny et al. [6, 7, 34] have shown, evaluating all candidate attribute-value combinations that qualify as QIDs, requires $\mathcal{O}(2^n)$ time, where n is the number of attributes. In large high dimensional datasets the scalability of the QID discovery algorithm is an issue because of the high number of attributes (e.g. $n \geq 100$). Having a QID discovery algorithm that is scalable to large high-dimensional datasets, can make a meaningful contribution is supporting big data anonymisation.
2. **Handling Dataset Changes.** In existing algorithms that rely on QID discovery to form anonymised datasets, changes to the data are not taken into consideration. That is they do not handle possible changes to the dataset due to the addition or deletion of attributes. For instance, if an anonymised dataset is integrated with some new data either deliberately or by accident; or what the impact of deleting certain attributes has on the privacy of the data. Discovering QIDs that emerge due to these changes in availability of information requires an efficient algorithm that is able to handle large dynamic datasets in order to effectively prevent re-identification attacks.

3. **Attack Model.** Existing mechanisms for preventing re- identification attacks on anoymised data work by devising a solution to specifically prevent a known re-identification attack from being conducted successfully on a dataset anonymised via a given procedure [15,18,39]. While this addresses known vulnerabilities, the limitation is that the strength of the anonymisation scheme is hinged on its ability to identify and prevent known disclosure patterns. Furthermore, this does not account for an adversary with knowledge of auxiliary data, the anonymisation mechanism, and high processing power.

In order to be effective at preserving privacy therefore, any QID discovery algorithm must account for an adversary that has knowledge of auxiliary data, the anonymisation mechanism, and high processing power.

Contributions. We make the following contributions in line with the problems outlined above.

1. **Find-QID.** In order to handle QID discovery on large high-dimensional datasets we propose a novel parallel search algorithm called *Find-QID* which operates by searching over attribute combinations in parallel instead of sequentially. The result is that it is faster and able to handle much larger datasets than state-of-the-art algorithms. Furthermore, *Find-QID* is adaptable to hardware processing architectures like Graphics Processing Units (GPUs) without loss in performance efficiency and can handle auxiliary data changes to the data without this adversely affecting performance.

2. **Scalability.** We propose a best-effort (greedy) search strategy that allows for an opportunistic search for minimum sized QIDs that are subsets of larger sized QIDs. Since these minimum QIDs are linkable to supersets of QIDs, it suffices to identify all the minimum QIDs to discover all the QIDs in the dataset. Thereby significantly reducing the time, storage, and memory requirements to support the search operation. This is, even as the dataset's size grows or shrinks, due to data modifications or deletions respectively.

3. **Preventing Re-Identification.** We propose an efficient validation mechanism based on an adversarial model that leverages the notion of discovering unique column combinations (candidate-keys) [2,29]. The goal is to demonstrate that failure by the adversarial model to discover candidate-keys indicates that re-identifying individual records based on QIDs is not possible on a dataset generated by *Find-QID*.

4. **Empirical Results.** We evaluate our solutions on a semi-synthetic health dataset compiled from various sources such as government websites and public statistical data. In terms of hardware, we benchmark the performance of *Find-QID* on two hardware architectures, namely: an NVIDIA Tesla V100 GPU, and an E5-2698V4 CPU machine. Our empirical results show that *Find-QID* not only scales well to large high-dimensional datasets, but exploits its parallelisability on GPU (Graphics Processing Unit) architectures to prevent re-identification even in the presence of a powerful adversary equipped with similar high-performance computing processing power.

Outline. The rest of the paper is structured as follows. In Sects. 2 and 3, we present our proposed *Find-QID* algorithm as well as our adversarial model. In Sect. 4, we present results from our empirical model to demonstrate the scalability of *Find-QID* and the adversarial model, on both a regular CPU powered machine and a GPU machine. We discuss related work in Sect. 5, and, conclude in Sect. 6 with a summary of our results as well as some suggestions for avenues for future work.

2   Discovering Quasi-identifiers

In this section we describe our quasi-identifier (QID) discovery scheme - *Find-QID*. We begin with a formal description of quasi-identifiers and then highlight why the QID discovery problem is challenging. To pave the way for the *Find-QID* scheme, we describe the Podlesny et al. [32,34] sequential (CPU) QID discovery scheme.

2.1   Quasi-identifiers

As mentioned before, the quasi-identifier (QID) discovery problem requires a solution that efficiently identifies attribute combinations that pose a re-identification threat (vulnerability). Typically, the individual attributes belonging to such QIDs are benign. The vulnerability emerges when the attributes are combined. A classic example was presented by Sweeney et al. [44,45] in 2002, by showing that 87% of the US population could be uniquely identified by combining the attributes: *ZIP, Birthday* and *Gender*. Therefore, any data set composed of more than 13% of the US population that contained a QID composed of a superset of attributes that included *ZIP, birthday* and *Gender* would be vulnerable to re-identification attacks based on QID values. However, as we explain in the following section, discovering all the QIDs in a given data set efficiently, is a challenging problem.

2.2   Enumeration and NP-Completeness

Bläsius et al. [6], demonstrated that the problem of detecting functional dependencies in a given dataset falls into the same class as the hitting set problem. Since the hitting set problem is NP-Complete and W[2]-Complete [6,19], by implication, the problem of functional dependency discovery is also NP-Complete and W[2]-Complete.

We note a correlation between the problem of detecting functional dependencies and that of discovering quasi-identifiers in that, both express dependencies between attribute values. In the case of a functional dependency say $F \rightarrow A$, the implication is that the values in the attribute A functionally depend on the values of F. For instance, a person's *Date-of-Birth* determines his/her *Age*. Likewise with QIDs, as mentioned before, a QID say *ZIP, Birthday* could determine a person's *Age* and *Residential Address*.

By analogy, therefore, the QID discovery problem falls into the same class as the Hitting Set Problem and that of detecting functional dependencies [5–7]. Therefore finding an efficient heuristic to address QID discovery on large high dimensional data sets can make an essential contribution to generating dependable (trustworthy) privacy-preserving data sets.

2.3   Characterising QIDs

We define a quasi-identifier (QID) as a relationship between attributes in a dataset (e.g. one that follows a relational schema) such that $Q \rightarrow_{ID} r_i$ where $Q = \{a_i...a_k\}$ is the combination of attributes that identifies a given record r_i [32]. A QID is valid (in that it successfully results in the re-identification of a record) for a record, r_i, if and only if for all records r in an equivalence class, E, associated with QID, Q, $r_i <> r$. In addition, the QID $Q \rightarrow_{ID} r_i$ is *non-trivial* if Q has not been already identified (e.g. with respect to another equivalence class E'), and Q is *minimal* if no QID $Q' \rightarrow_{ID} r_i$ exists such that $Q' \subset Q$ and a valid QID. To discover all QIDs therefore, it suffices to discover all minimal, non-trivial QIDs, because these QIDs allow for the identification of all QID supersets of the minimal QIDs. We discuss the process of discovering QIDs in the next Section.

2.4   Computing QIDs Sequentially

To compute the QIDs in a given dataset, Podlesny et al. [32,34] proposed a sequential algorithm based on the notion of discovering unique column combinations within a relational database [1,29]. The algorithm selects unique column combinations based on the *attribute value cardinality*. With attribute cardinality, a combination of attributes is classified as a QID (Unique Column Combination) if the number of unique values surpasses a pre-defined threshold. For example, if the threshold were set to a value say, k, all combinations appear "uniquely" less than k times.

More formally, the cardinality notion is defined as follows.

Definition 1. *Cardinality.*
The cardinality c of a combination of attributes is given by:

$$c = \frac{\text{Unique Values}}{\text{Total Number of Attribute Values}}$$

Since the cardinality is computed for all attribute-value combinations in the dataset, this operation is computationally intensive. In essence, the number of combinations that are evaluated is equivalent to:

$$\sum_{l=1}^{n} \binom{n}{l} = \sum_{l=1}^{n} \frac{n!}{l!(n-l)!} = 2^n - 1 \qquad (1)$$

where n is the size of attribute-value combination, and $n \geq 2$. Therefore, for very large high dimensional data sets (#Attributes $\geq$ 50), evaluating all attribute-value combinations ($n \geq 2$) for QID candidacy requires $\mathcal{O}(2^n)$ time [34].

Once all the QIDs that fall below the cardinality threshold have been identified, data privacy (anonymity) is enforced by using deterministic mechanisms such as generalisation, suppression, and perturbation to transform the data to minimise the risk of re-identification. Podlesny et al. propose the notion of compartmentation to handle this [34]. Compartmentation works in two steps. First, *data compartments* are created with attribute combinations that exclude the ones likely to cause re-identification and second the data within the compartments are transformed to ensure that no QIDs remain therein. For instance, for a set of attributes A, B, C, D, if B, C were classified as a QID. All compartment formations aim to avoid placing B, C together in any shared data subset. So, we might have the following combinations in valid safe disjoint compartments: $A, B, D, ...A, C, D, ...A, C, ...$. Therefore, subsets of the dataset can be shared safely without the risk of re-identification.

QID discovery however, is essential to ensuring privacy with data compartmentation and so it makes sense to seek a computationally efficient alternative. We propose such a solution in the next section.

2.5   The *Find-QID* Scheme

The *Find-QID* scheme addresses the issue of QID discovery with a parallel procedure based on the principle of *divide-and-conquer*. To conceptualise the operation of the *Find-QID* scheme, we employ a shared memory parallel computing model. This architecture was chosen to reflect a typical small scale parallel computing machine, based on the combination of a GPU (Graphics Processing Unit) and a CPU (Computer Processing Unit) to accelerate processing-intensive operations. We assume that the model is endowed with N CPU processors, that are connected to a single shared memory (e.g. a single motherboard and random access memory). The processors use the shared memory for the computations and operate synchronously, in that they all execute the same *Find-QID* algorithm on a subset of attributes assigned. Input and Output (I/O) from the processors is communicated based on the operations performed and the results generated.

For simplicity, and to support the description of the *Find-QID* scheme, we make the following assumptions regarding our shared memory model:

1. There exists a set of identical CPU processors $P_1, P_2, ..., P_N$ on the machine. Conceptually, N could be infinite, but for practical reasons we must assume that N is finite.
2. The shared memory unit has M memory locations, such that $M \geq N$.
3. A memory access unit handles (synchronises) processor access to memory. In the empirical model, this is handled by GPU that employs a randomised thread scheduling mechanism to allow the machine synchronise access to memory and the CPUs.

As mentioned before, since the sequential QID discovery algorithm requires $O(2^n)$ time, where n is the number of attributes, the *Find-QID* scheme aims to use this shared model to speed-up the computation time for QID discovery. As a point of note, in the empirical model (see Sect. 4, instead of traditional scalar based computation, the *Find-QID* scheme utilises vector and tensor-based calculation to increase processing efficiency (see Fig. 9). We now describe the *Find-QID* scheme designed for the aforementioned shared memory model.

Let $A = a_1, a_2, ..., a_n$ be a finite set of attributes in a dataset that is to be anonymised. The set of minimum-sized QIDs of A, denoted QIDs_A, describes all the smallest sized QIDs (minimum QIDs) which allow for all the QID supersets within the dataset to be discovered. Therefore, the problem is to compute QIDs_A from A efficiently, that is in time less than $O(2^n)$, but at the same time evaluate all combinations of a_i of lengths 2 to n. Quasi-identifiers of length 1 are known as standalone identifier and covered by related work for instances through their cardinality itself.

The *Find-QID* adopts a *multi-way divide and conquer* approach to do so. In the first step (Step 1), the cardinality of the attributes is computed in parallel following the approach described in Sect. 2.4. The attributes are then sorted according to their cardinality degree (level of entropy - number of unique values discovered). The second step (Step 2), involves partitioning A into $n - 2$ subsets such that we have $A_2, ..., A_i, ..., A_{n-1}$ where $A_i \subseteq A$ and A_i contains attribute combinations of size i. In the third step, (Step 3), *Find-QID* is applied recursively to all the A_i, in parallel (simultaneously).

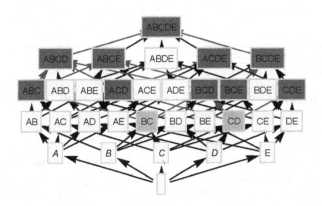

Fig. 1. Combination tree for QID candidates

Example 1. For example, for a four (5) attribute dataset with attributes (a, b, c, d, e), we have candidate combinations of size 2 to $n - 1 = 5 - 1 = 4$. So all candidate combinations of sizes 2, 3, and 4, are evaluated in parallel to determine which are QIDs. In the example given in Fig. 1 the following combinations of size 2:

$$\{AC, AD, AB, AE, BC, BD, BE, CD, CE, DE\}$$

ones of size 3:

$$\{ABC, ABD, ABE, ACD,$$

$$ACE, ADE, BCD, BCE, BDE, CDE\}$$

and ones of size 4:

$$\{ABCD, ABCE, ABDE, ACDE, BCDE\}$$

are evaluated for QID candidacy.

QID candidacy is determined based on whether or not the attribute values in the identified combinations contain functional dependencies that can result in disclosure. The cardinality metric plays a key role in this procedure, since uniqueness (expressed by a high cardinality) is indicative of the likelihood of the existence of functional dependencies between the attributes in the combination under evaluation. Papenbrock et al. offer an assessment of efficient algorithms developed for functional dependency discovery [30]. Yet, those algorithms are not directly applicable to QID discovery as QIDs answer to maximal partial unique column combinations (mpUCC), a special case of UCCs.

Finally, in the fourth step (Step 4), the sets of QIDs obtained in Step 3, are evaluated to obtain $QIDs_A$. The pseudo-code in Algorithm 1 summarises the *Find-QID scheme.* We now show how this step (Step 4) works.

Algorithm 1. *Find-QID* algorithm

1: **Step 1:** Attribute Cardinality
2: **for** $\{\forall a_i \;\&\&\; (1 \leq i \leq n)\}$ **do**   ▷ Do in parallel
3:     Split $|A|$ over P_N ▷ Split attributes over CPUs
4:     Sort a_i by cardinality degree

5: **Step 2:** Evaluate QID Combinations
6: **for** $\{\forall A_i \;\&\&\; (2 \leq i \leq n - 2)\}$ **do**   ▷ Do in parallel
7:     Split $|A_i|$ over P_N ▷ Split combinations over CPUs

8: **Step 3:** Recursive *Find-QID*
9: **for** $\{\forall A_i \;\&\&\; (2 \leq i \leq n - 2)\}$ **do**
10:     $Find\text{-}QID(A_i)$

11: **Step 4:** Form $QIDs_A$
12: **for** $\{\forall A_i \;\&\&\; (2 \leq i \leq n - 2)\}$ **do**
13:     $QIDs_{A_2} \leftarrow QIDs_{A_2} \ldots \bigcup QIDs_{A_i} \ldots \bigcup QIDs_{A_{n-2}}$
       return $QIDs_A$

2.6   Discovering QIDs

As mentioned before, it suffices to discover the set of minimal QIDs in order to obtain the set of all QIDs in a dataset that creates a re-identification vulnerability. To compute the set of minimal QIDs, each attribute combination of size i ($2 \leq i \leq n$) is evaluated to determine whether or not it is minimal. An attribute combination is minimal if it does not contain a subset of attributes that have already been classified as a QID.

We create a union of every two attributes in the first $i - 1$ columns (i is the size of the attribute combination) and then sort the discovered combinations lexicographically. Next, we crosscheck to ensure that the found combination is minimal and then check that it has not yet been discovered.

So essentially, for each attribute-value combination, we check to determine whether any of the existing (discovered minimal combinations) is a superset of the combination under consideration. If this is not the case, the combination is included in QIDsA. Example 2 depicts a case of computing QIDsA based on Fig. 1.

Example 2. If we consider the attribute combinations generated as permutations from $\{A, B, C, D, E\}$, obtaining QIDsA requires that each candidate be evaluated for minimality. Checking tuple length with size 2, one will discover that attribute tuples $\{BC, CD\}$ are QIDs. Now, looking at tuple length of 3, we know already that tuples $\{ABC, ACD, BCD, BCE, CDE\}$ are supersets of the minimal QIDs $\{BC, CD\}$ and consequently be QIDs as well making their processing redundant. Likewise applicable for $\{ABCD, ABCE, ACDE, BCDE\}$ for tuple length of 4 and $ABCDE$ as they are eliminated from the set of minimal QIDs. This implies that for tuple length of 3 it is required to **check just half of all combinations** namely $\{ABD, ABBE, ACE, ADE, BDE\}$, and for tuple length four only $\{ABDE\}$. So we finally have the following set of minimal QIDs: $\{BC, CD\}$ which can each be used to obtain A, B, C, D, E. In other words, it suffices to have any of $\{BC, CD\}$ in a dataset to successfully re-identify an individual.

The pseudo-code in Algorithm 2 summarises the selection of minimal QIDs as is done in Step 4 of the *Find-QID* scheme.

Algorithm 2. *Min-QID Select* algorithm

1: **for** $i = 2$ to $n - 2$ **do**
2:      Find minimum QIDs of size i
3:      **for** $\{\forall a_j \,\&\&\, (1 \leq j \leq i)\}$ **do**   ▷ Do in parallel
4:          $a_j \leftarrow a_1 \bigcup a_2 \bigcup ... a_i$
5:          **if** $\min(a_j)$ is TRUE **then**
6:              QIDs$_{A_j} \leftarrow a_j$
      return QID$_{A_j}$

2.7   Complexity Analysis - Discussion

We consider the best and worst case time complexity of the *Find-QID* scheme in order to evaluate the speed-up obtained by parallelising QID discovery. Let $T(n)$ be the worst-case running time of the sequential (CPU) QID discovery algorithm, so that $T(n) = O(2^n)$ as mentioned in Sect. 2.4. For *Find-QID*, let $T(p)$ be the worst case running time using p processors such that $p \leq N$, then $T(p) = O(\frac{2^n}{p})$.

To determine the speedup afforded by parallelism, in the best case if $p = 2^n$, the speed-up $S(1, p) = \frac{T(1)}{T(p)} = \frac{O(2^n)}{O(1)} = O(2^n)$. That is we have a speed-up in $O(2^n)$ which implies that the QIDs can be found in constant time using *Find-QID*, if there can be at least 2^n simultaneous compute operations executed on the machine. For the upcoming experiments in Sect. 4, the hardware offers 1120 trillion floating-point operations per second.

In the average case where $p \geq n$ but $n < 2^n$, $T(p) = O(\frac{2^n}{n})$ and so the speed-up is $S(1, p) = \frac{T(1)}{T(p)} = \frac{O(2^n)}{O(\frac{2^n}{n})} = O(n)$.

Finally, in the worst-case, where $p < n$ and p 1, $S(1, p) = O(p)$ implies that fewer processors negatively impacts on the performance of *Find-QID* as the cost of the thread scheduling operations out-weighs the benefit of parallelism.

3   Attack Model

We propose an efficient validation mechanism based on an attack model that leverages the notion of discovering unique column combinations (candidate-keys) from the data profiling domain. The goal is to demonstrate that failure by the attack model to find candidate-keys indicates that re-identifying individual records are not possible on a dataset generated by *Find-QID*.

The validation mechanism is based on a QID attack model that applies a greedy search for unique attribute values in a given dataset. Process-wise, the attack model works by iterating over attribute-value combination candidates to determine the conditional probability that the combination is likely to be useful for re-identification. So, if we assume that we have a relational database model, for each tuple, the attack model maintains a *candidate queue* ordered by likelihood of re-identification. Based on this queue, vector-based GROUP and COUNT operations are executed to determine the equivalence class size k for the designated candidate combinations. The attribute-value combination counts (numbers) that fall below k are then listed as "opportunities" for exploitation. Essentially, the attack model aims at achieving two (2) things through the process of forming the *candidate queue*:

- Identifying the highest possible number of attribute-value combinations that have a high conditional probability for re-identification for the **same record** (row) to increase the success of cross-linkage, and;
- Identifying the highest possible number of attribute-value combinations that have a high conditional probability for re-identification for **different distinct**

records (rows) to increase the information gain in terms of the number of different individuals identified.

The conditional probability is computed as the weighted sum of the cardinalities of the individual attributes. As a next step, the attack model uses the list of QIDs to serve as candidate-keys that can be combined with auxiliary data using a conventional JOIN operation to elevate the information gain. A high level of information gain increases the possibility of re-identifying individual tuples exactly using cross-linkage. The higher the conditional probability, the more likely an attribute tuples serves as QID and can be successfully exploited to cross-link the same record to auxiliary data.

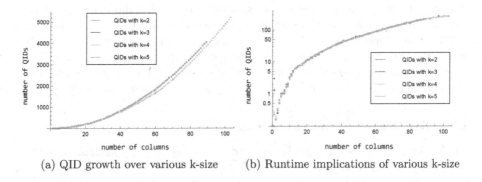

(a) QID growth over various k-size   (b) Runtime implications of various k-size

Fig. 2. QID characteristics

Finally, to increase the power of the attack model, we allow the attack model to randomly adjust (increase or decrease) the equivalence class size k as a further attempt to reveal unprocessed QIDs. The goal is to aggravate cross-linkage by growing the intersection of potential individuals with similar profiles to that in the QID with a high conditional probability. The same comes true to Sweeney's original k-anonymity for k-size equivalent classes [45]. Increased k condition steepens the requisite bar of QIDs and therefore decreases their amount (see Fig. 2a) while the runtime remains almost unaffected (see Fig. 2b). For Fig. 2a and Fig. 2b, k = 4 and k = 5 number of QIDs are overlapping with k = 3.

4   Empirical Model

In our empirical model, we aim to show that our proposed *Find-QID* scheme is scalable to large high-dimensional datasets and practically applicable to existing parallel computing hardware. We also show the performance gains of the *Find-QID* over the CPU-Compute scheme for quasi-identifier discovery. To this end, we employed the following hardware and datasets for our experiments.

Hardware. Our experiments were conducted on a GPU-accelerated high-performance compute cluster, housing 160 CPU cores (E5-2698 v4), 760 GB RAM, and 10x Tesla V100 with 5120 CUDA cores each and a combined Tensor performance of 1120 TFlops. The execution environment for GPU related experiments will be restricted to one dedicated CPU core and a single, dedicated Tesla V100 GPU. For CPU related experiments, the runtime environment on the compute cluster is bound to 10x dedicated cores.

Data Sets. To allow reproducible experiments and the disclosure of raw data samples for comparison, a semi-synthetic dataset was compiled. Various sources have been concatenated, including official government websites, public statistical data and datasets as part of previous publications. A complete list of all attributes is publicly available on github.com [31]. In sum, 110 describing attributes exists across one million rows, experiments will be conducted on all 110 attributes and a random sample of 100k rows. The dataset itself consists out of distribution-sensitive but fake names, SSN, phone numbers and addresses as well as disease, drugs and single-nucleotide polymorphism (SNP) records as genomic mutations to mimic a typical health dataset we have seen in hospitals.

4.1   QID Discovery and Processing

To identify attribute-value combinations that are unique, we implemented this by counting the conjoined presence for each attribute-value combination, using an SQL *GROUP BY* and *COUNT* operation over all the attribute combinations in the dataset. Attribute value combinations that appear less than k-times (where k is the threshold of the minimum number of similar records required to preserve anonymity [45]) qualify as *QIDs*. The identified *QIDs* are then transformed to remove the risk of re-identification.

Transforming the discovered *QIDs* is achieved by a combination of the following procedures: (1) *perturbation* [24,38]; (2) *generalisation* and *suppression* [22,25,45]. So, for our experiments, we say that we have an anonymised dataset once we have identified all the *QIDs* and transformed them to prevent re-identification. We use such anonymised data to simulate anonymisation efforts for the attack simulation later (see Sect. 4.5).

As mentioned before, the CPU-compute approach to *QID* discovery that we have just described is computationally intensive. In essence, searching and identifying QIDs is equivalent to finding attribute value subsets (of any length) which are unique or size k and identify designated rows. In the field of data profiling, those attribute value subsets are referred to maximal partial Unique Column Combinations (mpUCC), so UCCs which do not apply to all rows (UCC) but rather a subset (mpUCC) for the given case k. Traditionally, UCC discovery can be accomplished by some bottom-up or top-down principle [1,16,17].

In the following, we describe how the *Find-QID* scheme was implemented to work on the GPU hardware described above, and the heuristics used to start computation opportunistically at most promising candidates in the attribute-value combinatorial hierarchy (see Fig. 1).

4.2 Implementing *Find-QID*

We implemented the *Find-QID* scheme as a composition of four components namely: *Preprocessor*, *QID Selector*, *Compartmenter* and *Verifier*. Preprocessing and QID selector are engaged in an incremental fashion increasing over the number of tuples length for $2..n$ where n describes the maximum number of attributes in the dataset (see Fig. 3).

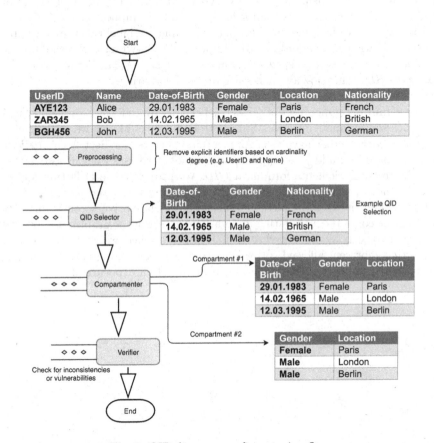

Fig. 3. QID discovery and processing flow

Preprocessing. In the first step, all attribute-value combinations for a specific length of L (conditional on the number of tuples) are generated. Each of these combinations is checked against the list of already discovered *QIDs* (attribute-value combinations that enable re-identification). In line with the goal of identifying the minimum-sized *QID* as means of discovering all *QID* supersets, we verify whether or not the *QID* in question has siblings or parents in the combination hierarchy (see Fig. 1). If this is the case, the attribute-value combination candidates that qualify as a parent of a (minimal) QID can be neglected from further processing to eliminate this combinatoric branch. We now consider how to select *QIDs*.

QID Selector. For the remaining attribute-value combination candidates that are not eliminated from further processing, we must determine their degree of uniqueness. This is used to decide to what level the combination can be exploited for re-identification. We do this by computing the count of rows (in parallel) that are unique by utilising a parallel *GROUP BY* and *COUNT* operation. So, if there is a group size of less than k for $k \leq 2$, which implies that at least one specific row is identified uniquely through this tuple, we label this the attribute-value combination as a candidate QID. Implementation-wise the threshold point k can be adjusted based on specific privacy needs (see Fig. 2a). We now consider how *Find-QID* forms safe anonymised datasets for sharing.

Compartmenter. Once a set of *QIDs* has been discovered, they can be now further processed for anonymisation. Similar to Podlesny et al. [34] we use *attribute compartmentation* to achieve this. As mentioned before (see Sect. 2.4), *attribute compartmentation* builds overlapping partitions of attributes without association of attribute tuple elements forming a *QID*. We project these QID tuples onto a graph schema, where the *attribute-values* represent the nodes and the edges, the connections between the *attribute-values* that violate privacy constraints. By inverting the edges (i.e. computing an independent set of the graph), we obtain a graph that consists of all allowed combinations. In essence, we have the maximal cliques that represent allowed (privacy-preserving) overlapping partitions which can be shared publicly without the risk of private data exposure [34].

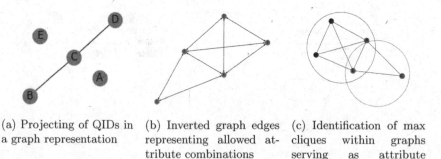

(a) Projecting of QIDs in a graph representation

(b) Inverted graph edges representing allowed attribute combinations

(c) Identification of max cliques within graphs serving as attribute compartments

Fig. 4. Quasi-identifier processing during attribute compartmentation

Verifier. Finally, to cross-check for privacy vulnerability and also in line with our attack model (see Sect. 3), we employ a verifier to test the attackability of the data generated by the *Compartmenter*. This step is optional and only necessary if parallelism computing threads fail and do not raise exceptions. Without existing unique attribute combinations, as each subset of record is now represented at least k times, mapping auxiliary data bijective is impossible and a projection can be done only in a probabilistic manner.

4.3   Parallelism

Computation heavy is especially the *preprocessing* and *QID selection* given the sheer amount of combinations. Through the incremental nature, all computation activities within one layer (same candidate tuple length) can be parallelised as their calculation do not depend upon each other. When pointing the reader's attention towards Fig. 1 for a second time, we see the levels of attribute combination length represented. At the top, the first tuple length is none, then one, two, three, and finally four element tuples. As there are no dependencies within a given layer, just in-between parent and siblings, such nature can be utilised for parallelised task execution. In particular, each candidate within the same level can be parallelised in best case considering chunking to reduce process spawn efforts. Both, *preprocessing* and *QID Selector* require to be executed sequentially (iterative) with incrementing levels (L). Yet, the compute-heavy tasks within a specific level can be parallelised without any risk of dependencies. And after each increment of L, *preprocessing* will be applied again to filter already found minimal QIDs and significantly reduce candidates in the next iteration. To determine cardinality algorithmic, we employ the SQL operations of: *GROUP BY* and *COUNT*. Grouping by the attribute combination considered as a candidate and counting the matching rows returns the desired equivalent class k. As this basic operation is supported in almost any universe, we can easily utilise out-of-the-box libraries like the open data science framework cuDF[1], a RAPIDS Nvidia initiative which enables GPU acceleration. Using chunking, the grouping activity is rendered as vector operation [27, 28, 40] rather than a scalar one potentially even on multi-GPU environments. Chunking avoids overhead in process spawning as multiple tasks are bundled as bulk load and passed along together.   Within each chunk, compute, and memory is shared and spread across blocks with many threads operating independently and simultaneously (see Fig. 9). The implication of this will be displayed and discussed in the empirical analysis (Sect. 4). To offer a baseline for comparison, we have added data points using a published parallel QID search scheme by Braghin et al. [8].

4.4   Performance Comparison on GPU vs. CPU Architectures

Purely relying on vertical scaling by adding more CPU cores may look promising, yet Fig. 5a demonstrates just little relief for the actual runtime. Therefore, to benchmark the efficiency uplift of different architectures, Fig. 5b delineates the growth of execution time over the number of columns. The more describing attributes are available and require processing, the higher its execution time.

---

[1] https://github.com/rapidsai/cudf.

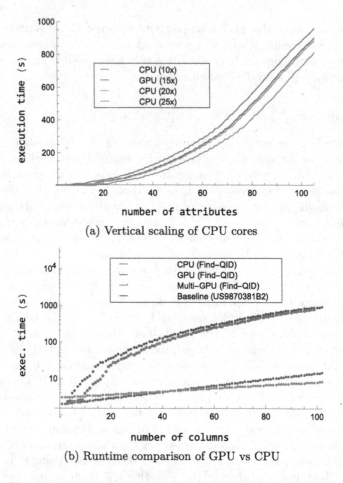

(a) Vertical scaling of CPU cores

(b) Runtime comparison of GPU vs CPU

Fig. 5. Execution characteristics over runtime and #columnss

Utilising parallelisation, the exponential nature in growth can be delayed in both architectures. Yet particularly in the GPU context through its capabilities of massive parallelisation the increase in runtime remains sustainable even for large-scale (see Fig. 5b). There are, however, situations especially in smaller dimensions where shifting to GPUs is counterproductive given the initial overhead of memory shifting through the motherboard bus.

Varying Number of Records. As part of the experiments, we have varied the size of rows increasing from 100k to 1M. Both CPU and GPU wise, the runtime implication was minimal as the combinatoric complexity did not change (see Fig. 6). The number of records, however, heavily influenced the required memory allocation, and especially GPUs memory is currently still limited to 16 GB to 32 GB without swapping to CPU memory.

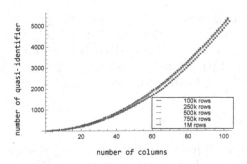

Fig. 6. GPU QIDs over rows and columns

Varying Number of Attributes. In contrast to the rows, the number of attributes substantially influenced the runtime as with increasing numbers of attributes and therefore columns the combination possibilities grow and consequently the processed tuples as well.

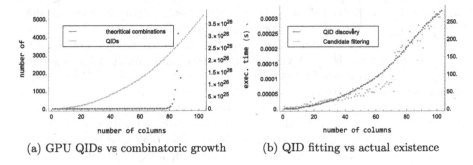

(a) GPU QIDs vs combinatoric growth   (b) QID fitting vs actual existence

Fig. 7. GPU experiments

Figure 7a delineates the growth of theoretical attribute combinations required processing against the QIDs found. The reader can observe the exponential nature of the increase.

Varying Cluster Sizes. The equivalence class (cluster) of size k is defined in a manner similar to k-anonymity the least number of identical data records for the same attribute value. For $k = 2$, at least 2 distinct rows should have exactly the same subset of attribute values. By adjusting this setting, the balance between information loss and equivalence class can be weighted. This is delineated in Fig. 2a through the different growth of QIDs given different designated k's. Here, the larger k, the more rows are required to qualify as QID and therefore – having a higher QID requirement – fewer QIDs are labelled in the same dataset.

GPU-Based Greedy QID Discovery. While the worst-case runtime remains the same, having smart decisions on the starting points of a greedy QID discovery can make a significant difference. Figure 7b depicts the runtime of both the *preprocessoring* and *QID Selector* activities over the growth in columns. Both are expectedly increasing in execution time. Due to the preprocessing, the QID discovery's actual runtime remains impressively low even for large attribute numbers (100+) and rows (1M). At this point, kindly highlighting that 100 columns require validating $2^{100} = 1.2 * 10^{30}$ attribute combinations.

4.5   Attack Model Evaluation

In our first experiment, with our attack model, we used a Twitter dataset containing about 500.000 random, anonymised tweets. We employed the attack model described in Sect. 3. In this case, we were able to re-identify up to 30% of the existing user base with 1:1 exact matches. This was achieved, on large-scale by analysing textual similarities like the word count within the tweet text in combination with, for instance, the user location. All of these attribute-value combinations form unique attribute values that serve as quasi-identifiers (QIDs). The dataset of 500.000 rows and 31 describing attributes contained 205 distinct QIDs.

After discovering QIDs in the described dataset, our second experiment with our attack model involved eliminating the same (identified) attribute-value combinations that made re-identification possible in the Twitter dataset. The identified combinations were eliminated using the *Find-QID* scheme described in Sect. 2.5. In other words, the Twitter dataset is now anonymised and our experiment is to demonstrate that it is secure to re-identification attacks. The resulting anonymised Twitter dataset was then subjected to the following attack.

Using the high-performance GPUs hardware architecture described above, we consider a scenario in which the attacker is endowed with the benefit of significant processing power similar to or better than that used to anonymised the data. The attacker's speed is boosted through massive parallelisation of computation heavy tasks. Figure 8a depicts the compute capacity with respect to the execution time, and Fig. 8b shows the memory utilisation. Both Figures are aiming at illustrating the capacity usage for both compute and memory resources. In this case, we note an average of 20% for computing, and 80% for memory resources. These experiments indicate the speed with which the attacker could potentially de-anonymise the dataset using our described attack model.

Based on this processing and memory resource usage, we tested our attack model on the anonymised Twitter dataset and obtained that not a single QID could be found. Lacking QIDs, we note that an exact cross-link to auxiliary data is not possible and the attack model successfully failed to ensure anonymity to the best of our knowledge.

5   Related Work

Several works have studied the issue of how to anonymise data sets in ways that provide firm guarantees of privacy [22, 25, 43–45, 50]. One of the challenges

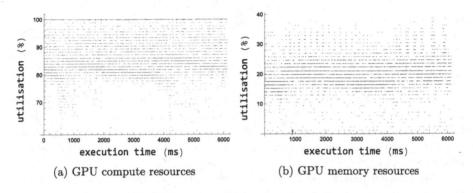

(a) GPU compute resources   (b) GPU memory resources

Fig. 8. GPU utilisation over execution time

of generating and sharing privacy-preserving dataset however, is that of anonymising large high-dimensional data sets effectively [4]. Large high dimensional datasets typically contain a high number of attributes but are usually sparsely populated in terms of attribute values. Most existing anonymisation mechanisms fail at generating privacy-preserving high dimensional data, because the conditions for creating safe equivalence classes based on grouping records by quasi-identifier similarity cannot be applied successfully without high information loss [11–13, 21, 22, 46–48]. High information loss renders the data useless for data analytics operations.

In recent work, Podlesny et al. [32, 34, 37] proposed addressing the issue of anonymising high dimensional data, by identifying all the minimal quasi-identifiers in a given dataset. Discovering the set of minimal quasi-identifiers guarantees that all supersets of quasi-identifiers in the dataset will be discovered as well. Once the quasi-identifiers are found, data compartments are created to achieve two things: (1) Form equivalence classes on the basis of quasi-identifier similarity; (2) prevent record re-identification by creating data compartments that exclude the attributes that make re-identification possible.

Discovering quasi-identifiers efficiently is however a challenging problem [35, 36]. While Podlesny et al. [33, 35] have proposed heuristics to address the issue, the worst-case time complexity remains in $O(2^n)$ [6, 32].

Parallel quasi-identifier discovery is one method of addressing the time complexity issue of the sequential quasi-identifier discovery algorithm. However, existing schemes have focused more on heuristics to minimise information loss and circumvent re-identification attacks rather than on performance speed-up.

Other related venues include Motwani et al. early work on measures for quantifying quasi-identifiers through distinct- and separation ratio [26], and Zare-Mirakabad et al. contribution of linking l-diversity diagnosis as a knowledge discovery problem of association rule mining to the data mining field [49]. Li et al. further extended the perspective on association rule mining and k-item sets by implementing a MapReduce approach as a parallel Apriori algorithm [23]. Traditional MapReduce technologies distribute the workload across multiple processing nodes thence introduce the drawback of slow network I/O. In the recent past, we were able to observe a growing availability of cheaper high-performance memory. Therefore, considering parallel algorithms for quasi-identifier discovery in a vertical setting becomes a suitable and practical choice again for privacy-preserving data generation.

Combinatorial optimisation problems such as quasi-identifier discovery have been shown to benefit from the GPU thread scheduling process [9,27,28]. The GPU thread scheduling process is well-adapted to running parallel computing algorithms that require a complete listing of all items in a collection (enumeration problems) [9,27,28].

This massive parallelism does not serve every use case but is particularly useful for solving combinatorial optimisation problems such as QID discovery. GPUs are receiving increasing attention as a hardware platform for supporting compute-intensive enumeration problems that typically require ordering results or values, and involve evaluating a high number of cases time (as well as computationally) efficiently.

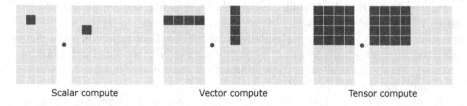

Scalar compute   Vector compute   Tensor compute

Fig. 9. Compute primative differences between CPU (scalar), GPU (vector) and Tensor (tensor)

Large L1-L3 caches characterise traditional CPU architecture, low compute density, low latency tolerance and optimised for serial scalar operations with a wide range of instruction set. By contrast, as illustrated in Figure 9 the GPU architecture is extremely customised primarily for vector-based operations with selective instruction sets, high compute density, high throughput, high latency tolerance and high computations per memory access [27,28].

As we have shown in Sects. 2, 3 and 4, parallelism is useful not only in creating privacy-preserving large high dimensional data sets, but also effective in preventing re-identification attacks either based on inferences drawn solely from the data set at hand or from associations made with auxiliary datasets.

6   Conclusions and Future Work

Finding attribute combinations that qualify as quasi-identifiers (QIDs) is an integral part of the process of generating privacy-preserving datasets to support analytics operations such as privacy-preserving data mining (or machine learning). Oftentimes however, the usability of privacy-preserving datasets is impeded either by it vulnerability to re-identification attacks or high-degree of information loss. In either case, the data is not usable for the purpose for which it was generated, which defeats the purpose of creating such datasets in the first place.

In this paper, we pointed out that *QID* discovery on large high-dimensional datasets, poses a further challenge. This is because the sparsity of the data in comparison to the number of attributes causes a high-level of information loss when similarity clusters (guided by cluster-size thresholds) are formed to anonymise the data. High-information loss renders the data useless for analytics operations. Podlesny et al. [32,34] proposed a method of identifying *QIDs* on large high-dimensional datasets. However, because this approach is CPU-compute oriented, it does not handle large high-dimensional datasets efficiently.

For that purpose, we presented *Find-QID* as a parallel (GPU-compute) approach to QID discovery. The *Find-QID* scheme handles *QID* discovery on large scale, high-dimensional datasets efficiently. This, in spite of the fact that *QID* discovery requires $\mathcal{O}(2^n)$ compute time and is NP-hard as well as W[2]-complete nature. We showed that using *Find-QID* we can achieve a computational processing speed-up of $O(p)$ in the worse case, and that no re-identifying records remain for any given an arbitrary similarity cluster of size k. Our attack model shows that this in turn prohibits various cross-linkage attacks like with auxiliary data to re-identify individuals as no unique attribute values of any length remain. As such *Find-QID* offers the opportunity of anonymising and validating datasets for privacy compliance.

A key result of our empirical model is that with massive parallelism we can achieve near real-time processing of huge datasets (e.g., $n = 100$ for n answering to describing attributes) without compromising privacy while some of those application fields have been declared as impractical until now [14].

As future work, it would be interesting to study the possibility of GPU acceleration for supporting generating privacy-preserving datasets in real-time. As part of the experiments, we point out that the full potential of the machine is not exploited despite the excellent runtimes (see Fig. 8a). Therefore, developing mechanisms to optimise memory swapping, and pre-fetching might result in further performance improvements (we estimate another 20% to 40%). Other work might involve considering the typical pitfalls in GPU accelerations, or CPU hybrid models, where data I/O might degrade processing time (more compute time) for smaller datasets.

References

1. Abedjan, Z., Naumann, F.: Advancing the discovery of unique column combinations. In: Proceedings of the 20th ACM International Conference on Information and Knowledge Management, pp. 1565–1570 (2011)
2. Abedjan, Z., Golab, L., Naumann, F.: Profiling relational data: a survey. VLDB J. **24**(4), 557–581 (2015). https://doi.org/10.1007/s00778-015-0389-y
3. Abedjan, Z., Golab, L., Naumann, F., Papenbrock, T.: Data profiling. Synth. Lect. Data Manage. **10**(4), 1–154 (2018)

4. Aggarwal, G., et al.: Anonymizing tables. In: Eiter, T., Libkin, L. (eds.) ICDT 2005. LNCS, vol. 3363, pp. 246–258. Springer, Heidelberg (2004). https://doi.org/10.1007/978-3-540-30570-5_17

5. Birnick, J., Bläsius, T., Friedrich, T., Naumann, F., Papenbrock, T., Schirneck, M.: Hitting set enumeration with partial information for unique column combination discovery. In: Proceedings of the VLDB Endowment vol. 13, no. 11, pp. 2270–2283 (2020)

6. Bläsius, T., Friedrich, T., Schirneck, M.: The parameterized complexity of dependency detection in relational databases. In: Guo, J., Hermelin, D. (eds.) 11th International Symposium on Parameterized and Exact Computation (IPEC 2016), volume 63 of Leibniz International Proceedings in Informatics (LIPIcs), pp. 6:1–6:13, Dagstuhl, Germany. Schloss Dagstuhl-Leibniz-Zentrum fuer Informatik. ISBN: 978-3-95977-023-1 (2017). https://doi.org/10.4230/LIPIcs.IPEC.2016.6, http://drops.dagstuhl.de/opus/volltexte/2017/6920

7. Bläsius, T., Friedrich, T., Lischeid, J., Meeks, K., Schirneck, M.: Efficiently enumerating hitting sets of hypergraphs arising in data profiling. In: Algorithm Engineering and Experiments (ALENEX), pp. 130–143 (2019)

8. Braghin, S., Gkoulalas-Divanis, A., Wurst, M.: Detecting quasi-identifiers in datasets. US Patent 9,870,381, 16 January 2018

9. Cook, C., Zhao, H., Sato, T., Hiromoto, M., Tan, S.X.-D.: GPU-based ising computing for solving max-cut combinatorial optimization problems. Integration 69, 335–344. ISSN: 0167-9260 (2019). https://doi.org/10.1016/j.vlsi.2019.07.003, http://www.sciencedirect.com/science/article/pii/S0167926019301348

10. Heer, D., Podlesny, J.: Process for the user-related answering of customer inquiries in data networks. US Patent 10,033,705, 24 July 2018

11. Dwork, C.: Differential privacy: a survey of results. In: Agrawal, M., Du, D., Duan, Z., Li, A. (eds.) TAMC 2008. LNCS, vol. 4978, pp. 1–19. Springer, Heidelberg (2008). https://doi.org/10.1007/978-3-540-79228-4_1

12. Dwork, C.: Differential privacy. In: van Tilborg, H.C.A., Jajodia, S. (eds.) Encyclopedia of Cryptography and Security. Springer, Boston (2011). https://doi.org/10.1007/978-1-4419-5906-5_752

13. Dwork, C., Roth, A., et al.: The algorithmic foundations of differential privacy. Found. Trends® Theoret. Comput. Sci. 9(3–4), 211–407 (2014)

14. Gutmann, A., et al.: Privacy and progress in whole genome sequencing. Presidential Committee for the Study of Bioethical (2012)

15. Hamza, N., Hefny, H.A., et al.: Attacks on anonymization-based privacy-preserving: a survey for data mining and data publishing (2013)

16. Han, S., Cai, X., Wang, C., Zhang, H., Wen, Y.: Discovery of unique column combinations with hadoop. In: Chen, L., Jia, Y., Sellis, T., Liu, G. (eds.) APWeb 2014. LNCS, vol. 8709, pp. 533–541. Springer, Cham (2014). https://doi.org/10.1007/978-3-319-11116-2_49

17. Heise, A., Quiané-Ruiz, J.-A., Abedjan, Z., Jentzsch, A., Naumann, F.: Scalable discovery of unique column combinations. Proc. VLDB Endowment 7(4), 301–312 (2013)

18. Ilavarasi, A.K., Sathiyabhama, B., Poorani, S.: A survey on privacy preserving data mining techniques. Int. J. Comput. Sci. Bus. Inform. 7(1) (2013)

19. Karp, R.M.: Reducibility among combinatorial problems. In: Miller, R.E., Thatcher, J.W., Bohlinger, J.D. (eds.) Complexity of Computer Computations. IRSS, pp. 85–103. Springer, Boston (1972). https://doi.org/10.1007/978-1-4684-2001-2_9

20. Kavitha, S., Yamini, S., et al.: An evaluation on big data generalization using k-anonymity algorithm on cloud. In: 2015 IEEE 9th International Conference on Intelligent Systems and Control (ISCO), pp. 1–5. IEEE (2015)
21. Kushida, C.A., Nichols, D.A., Jadrnicek, R., Miller, R., Walsh, J.K., Griffin, K.: Strategies for de-identification and anonymization of electronic health record data for use in multicenter research studies. Med. Care **50**, S82–S101 (2012)
22. Li, N., Li, T., Venkatasubramanian, S.: t-closeness: privacy beyond k-anonymity and l-diversity. In: 2007 IEEE 23rd ICDE, pp. 106–115, April 2007. https://doi.org/10.1109/ICDE.2007.367856
23. Li, N., Zeng, L., He, Q., Shi, Z.: Parallel implementation of apriori algorithm based on mapreduce. In 2012 13th ACIS International Conference on Software Engineering, Artificial Intelligence, Networking and Parallel/Distributed Computing, pp. 236–241. IEEE (2012)
24. Liu, K., Kargupta, H., Ryan, J.: Random projection-based multiplicative data perturbation for privacy preserving distributed data mining. IEEE Trans. Knowl. Data Eng. **18**(1), 92–106 (2006)
25. Machanavajjhala, A., Kifer, D., Gehrke, J., Venkitasubramaniam, M.: l-diversity: privacy beyond k-anonymity. ACM TKDD **1**(1), 3 (2007)
26. Motwani, R., Xu, Y.: Efficient algorithms for masking and finding quasi-identifiers. In: Proceedings of the Conference on Very Large Data Bases (VLDB), pp. 83–93 (2007)
27. Nickolls, J., Dally, W.J.: The GPU computing era. IEEE Micro **30**(2), 56–69 (2010)
28. Owens, J.D., Houston, M., Luebke, D., Green, S., Stone, J.E., Phillips, J.C.: GPU computing. Proc. IEEE **96**(5), 879–899 (2008)
29. Papenbrock, T., Naumann, F.: A hybrid approach for efficient unique column combination discovery. Technologie und Web (BTW), Datenbanksysteme für Business, p. 2017 (2017)
30. Papenbrock, T., et al.: Functional dependency discovery: an experimental evaluation of seven algorithms. Proc. VLDB Endowment **8**(10), 1082–1093 (2015)
31. Podlesny, N.J.: Semi-synthetic genome data (2020). https://github.com/jaSunny/synthetic_genome_data
32. Podlesny, N.J., Kayem, A.V.D.M., von Schorlemer, S., Uflacker, M.: Minimising information loss on anonymised high dimensional data with greedy in-memory processing. In: Hartmann, S., Ma, H., Hameurlain, A., Pernul, G., Wagner, R.R. (eds.) DEXA 2018. LNCS, vol. 11029, pp. 85–100. Springer, Cham (2018). https://doi.org/10.1007/978-3-319-98809-2_6
33. Podlesny, N.J., Kayem, A.V.D.M., Meinel, C.: Identifying data exposure across high-dimensional health data silos through Bayesian networks optimised by multi-grid and manifold. In: IEEE 17th International Conference on Dependable. Autonomic and Secure Computing (DASC), p. 2019. IEEE (2019)
34. Podlesny, N.J., Kayem, A.V.D.M., Meinel, C.: Attribute compartmentation and greedy UCC discovery for high-dimensional data anonymization. In: Proceedings of the Ninth ACM Conference on Data and Application Security and Privacy, pp. 109–119. ACM (2019)
35. Podlesny, N.J., Kayem, A.V.D.M., Meinel, C.: Towards identifying de-anonymisation risks in distributed health data silos. In: Hartmann, S., Küng, J., Chakravarthy, S., Anderst-Kotsis, G., Tjoa, A.M., Khalil, I. (eds.) DEXA 2019. LNCS, vol. 11706, pp. 33–43. Springer, Cham (2019). https://doi.org/10.1007/978-3-030-27615-7_3

36. Podlesny, N.J., Kayem, A.V.D.M., Meinel, C.: How data anonymisation techniques influence disease triage in digital health: a study on base rate neglect. In: Proceedings of the 2019 International Conference on Digital Health. ACM (2019)

37. Podlesny, N.J.: High-dimensional data anonymization for in-memory applications. US Patent 10,747,901, 18 August 2020

38. Polat, H., Du, W.: Privacy-preserving collaborative filtering using randomized perturbation techniques. In Third IEEE International Conference on Data Mining. ICDM 2003, pp. 625–628. IEEE (2003)

39. Presswala, F., Thakkar, A., Bhatt, N.: Survey on anonymization in privacy preserving data mining (2015)

40. Sanders, J., Kandrot, E.: CUDA by Example: An Introduction to General-Purpose GPU Programming. Addison-Wesley Professional, Boston (2010)

41. Sopaoglu, U., Abul, O.: A top-down k-anonymization implementation for apache spark. In 2017 IEEE International Conference On Big Data (Big Data), pp. 4513–4521. IEEE (2017)

42. Sowmya, Y., Nagaratna, M.: Parallelizing k-anonymity algorithm for privacy preserving knowledge discovery from big data. Int. J. Appl. Eng. Res. $11(2)$, 1314–1321 (2016)

43. Sweeney, L.: Simple demographics often identify people uniquely. Technical Report Working Paper 3, Carnegie Mellon University, USA (2000). https://projects.iq.harvard.edu/files/privacytools/files/paper1.pdf

44. Sweeney, L.: Uniqueness of simple demographics in the us population. LIDAP-WP4 (2000)

45. Sweeney, L.: Achieving k-anonymity privacy protection using generalization and suppression. Int. J. Uncertainty Fuzziness Knowl. Based Syst. $10(05)$, 571–588 (2002)

46. Wong, R.C.-W., Fu, A.W.-C., Wang, K., Pei, J.: Minimality attack in privacy preserving data publishing. In: Proceedings of the 33rd International Conference on Very Large Data Bases, VLDB 2007, pp. 543–554. VLDB Endowment. ISBN: 978-1-59-593649-3 (2007)

47. Wong, R.C.-W., Fu, A.W.-C., Wang, K., Pei, J.: Anonymization-based attacks in privacy-preserving data publishing. ACM Trans. Database Syst. $34(2)$. ISSN: 0362-5915 (2009). https://doi.org/10.1145/1538909.1538910

48. Wong, R.C.-W., Fu, A.W.-C., Wang, K., Yu, P.S., Pei, J.: Can the utility of anonymized data be used for privacy breaches? ACM Trans. Knowl. Discov. Data $5(3)$. ISSN: 1556-4681 (2011). https://doi.org/10.1145/1993077.1993080

49. Zare-Mirakabad, M.-R., Jantan, A., Bressan, S.: Privacy risk diagnosis: mining l-Diversity. In: Chen, L., Liu, C., Liu, Q., Deng, K. (eds.) DASFAA 2009. LNCS, vol. 5667, pp. 216–230. Springer, Heidelberg (2009). https://doi.org/10.1007/978-3-642-04205-8_19

50. Zhang, B., Dave, V., Mohammed, N., Al Hasan, M.: Feature selection for classification under anonymity constraint. arXiv preprint arXiv:1512.07158 (2015)

51. Zhang, X., Qi, L., He, Q., Dou, W.: Scalable iterative implementation of Mondrian for big data multidimensional anonymisation. In: Wang, G., Ray, I., Alcaraz Calero, J.M., Thampi, S.M. (eds.) SpaCCS 2016. LNCS, vol. 10067, pp. 311–320. Springer, Cham (2016). https://doi.org/10.1007/978-3-319-49145-5_31

52. Zimmermann, T., et al.: Detecting fraudulent advertisements on a large e-commerce platform. In: EDBT/ICDT Workshops (2017)

Towards Symbolic Time Series Representation Improved by Kernel Density Estimators

Matej Kloska[1]($\boxtimes$) and Viera Rozinajova[2]

[1] Faculty of Informatics and Information Technologies, Slovak University
of Technology in Bratislava, Ilkovičova 2, 842 16 Bratislava, Slovakia
`matej.kloska@stuba.sk`
[2] Kempelen Institute of Intelligent Technologies, Mlynské Nivy II.18890/5,
811 09 Bratislava, Slovakia
`viera.rozinajova@kinit.sk`
`https://www.fiit.stuba.sk`
`https://www.kinit.sk`

Abstract. This paper deals with symbolic time series representation. It builds up on the popular mapping technique Symbolic Aggregate approXimation algorithm (SAX), which is extensively utilized in sequence classification, pattern mining, anomaly detection, time series indexing and other data mining tasks. However, the disadvantage of this method is, that it works reliably only for time series with Gaussian-like distribution. In our previous work (Kloska and Rozinajova, dwSAX, 2020) we have proposed an improvement of SAX, called dwSAX, which can deal with Gaussian as well as non-Gaussian data distribution. Recently we have made further progress in our solution - edwSAX. Our goal was to optimally cover the information space by means of sufficient alphabet utilization; and to satisfy lower bounding criterion as tight as possible. We describe here our approach, including evaluation on commonly employed tasks such as time series reconstruction error and Euclidean distance lower bounding with promising improvements over SAX.

Keywords: Time series · Kernel density estimator · SAX · Tightness of lower bound

1 Introduction

It is generally known that huge amounts of data are generated on a daily basis, whereas streaming data make up a large part of them. There are many fields, such as healthcare, finance, security, energy and industry, where intelligent analysis and data mining tasks play an important role. Frequently the data capture event observations taken according to the order of time - we usually talk about time series. As we often deal with continuous data stream, it is clear, that its representation poses a significant problem when processing them.

© Springer-Verlag GmbH Germany, part of Springer Nature 2021
A. Hameurlain and A. M. Tjoa (Eds.): TLDKS L, LNCS 12930, pp. 25–45, 2021.
https://doi.org/10.1007/978-3-662-64553-6_2

At the same time, time series usually capture feature rich, highly dimensional data which make processing tasks even harder. In our work, the high dimensionality is related to a high number of varying data points in time series often capturing time series over a long period of time covering different behavior trends such as different power consumption during day and night, lockdown periods or seasonal differences. All these trends contribute to highly dimensional time series—which are difficult to process while capturing all significant properties over time.

Dimensionality reduction and descriptive forms of time series representation are recognized as a possible solution for highly performing data mining tasks [8]. A challenging area in the field of effective time series processing is their compact data presentation without sacrificing any significant information [23]. Symbolic representation of time series appears to be the solution to this problem giving us the opportunity to exploit longer time series periods without requiring significantly higher computational resources for most of data mining tasks.

The Symbolic Aggregate approXimation algorithm (SAX) [11] is one of the most popular symbolic mapping techniques for time series. SAX as a powerful unsupervised symbolic mapping technique is widely used due to its data adaptability. It is extensively utilized in sequence classification [17], pattern mining [4], anomaly detection [9] and many other data mining tasks [12,19,21]. However, this approach heavily relies on assumption that processed time series have Gaussian-like distribution [11]. When time series distribution is non-Gaussian or skews over time, this method does not provide sufficient symbolic representation.

In our previous work [10] we have proposed an improvement of SAX, called dwSAX, which can deal with Gaussian as well as non-Gaussian data distribution. Later, the similar approach was presented by Bountrogiannis et al. [2]. In this paper we introduce an improved technique for symbol breakpoints and centroids selection which contributes to more efficient alphabet symbols utilization. The goal is to optimally cover the information space and prove that our extension satisfies lower bounding criterion for wide exploitation of our method. The method was evaluated on commonly employed tasks such as time series reconstruction and Euclidean distance lower bounding with promising improvements over SAX.

This paper is organized as follows. Section 2 describes original SAX method and techniques for data distribution estimation. Section 3 introduces dwSAX - our extension to SAX method and its further improvement, which represents the core of this paper—edwSAX. Section 4 contains an experimental evaluation of the proposed method on time series reconstruction error and tightness of lower bound tasks compared to the original SAX method. Finally, Sect. 5 offers some conclusions and suggestions for future work.

2   Related Work

One of efficient data stream processing problems is their high dimensionality, too high number of data points due to various reasons such as mixed observed behaviors or gradual change over time. Possible solution to this issue is efficient symbolic representation of a high-dimensional data stream through a reduced dimensional symbolic data stream. In past decades, many different time series representations have been introduced. Lin et al. [11] divided methods into data adaptive (e.g. Piecewise Linear Approximation, Singular Value Decomposition, SAX) and non data adaptive (e.g. Wavelets, Random Mappings, Discrete Fourier Transformation). Recent research [3,13,14,21,24] shows activities in both method families. In the following sections we discuss fundamentals of the original SAX, and techniques for data distribution estimation.

2.1 Symbolic Representation - SAX

SAX is one of the best known algorithms for symbolic time series representation. This method makes it possible to represent any time series of length n using a string of any length w ($w \ll n$) with symbols from a predefined alphabet. The mentioned method consists of:

1. *dimensionality reduction*: applying Piecewise Aggregate Approximation (PAA) to the time series [8], this significantly reduces dimensionality and preprocesses time series for a further step;
2. *discretization*: mapping PAA segments into specific symbols from the alphabet based on a pre-computed mapping symbols table.

The concept of this method is illustrated in Fig. 1. Throughout this paper we use common notation used also in the original SAX paper [11] which is found Table 1.

Table 1. A summarization of common notation used in this paper and the original SAX paper [11].

C	A time series $C = c_1, ..., c_n$ where $c_i \in \mathbb{R}$
$\bar{C}$	A Piecewise Aggregate Approximation of a time series $\bar{C} = \bar{c}_1, ..., \bar{c}_w$
$\hat{C}$	A symbol representation of a time series $\hat{C} = \hat{c}_1, ..., \hat{c}_w$
w	The number of PAA segments representing time series C
a	Alphabet size (e.g., for the alphabet = {a, b, c}, $a = 3$)

Dimensionality Reduction: The intuition based on the aforementioned description is to reduce time series from n dimensions into w dimensions. This goal is simply achieved by division of the time series into w equal sized pieces - segments. For each piece, its mean value is calculated, and this value represents the underlying vector of w original values. Total vector of all pieces becomes a new reduced representation of the original time series.

More formally, a time series C of length n can be reduced into a w-dimensional time series by a vector $\bar{C} = \bar{c}_1, ..., \bar{c}_w$ where i^{th} element of $\bar{C}$ is calculated as follows [11]:

$$\bar{c}_i = \frac{w}{n} \sum_{j=\frac{n}{w}(i-1)+1}^{\frac{n}{w}i} c_j \qquad (1)$$

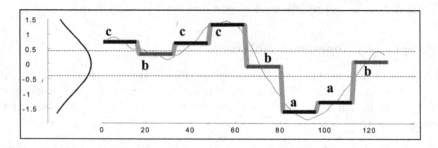

Fig. 1. Concept of original SAX. Background grey thin line is replaced with bold line segments (PAA). PAA segments are mapped by normal distribution sketched on y axis into symbols, $a = 3$. The symbols ranges are: $a = (-\infty; -0.43\rangle, b = (-0.43; 0.43\rangle, c = (0.43, \infty)$ [11].

Discretization: Discretization step replaces PAA segments obtained in the length reduction step by alphabet symbols. Assuming data are normalized before reduction (with zero mean), and have highly Gaussian-like distribution, the replacement is performed as follows. We first precompute a table of breakpoints. Lin et al. [11] defined breakpoints as a sorted list of numbers $B = \beta_1, ..., \beta_{a-1}$ such that the area under a $N(0,1)$ Gaussian curve from β_i to $\beta_{i+1} = 1/a$ (β_0 and β_a are defined as $-\infty$ and ∞, respectively).

The Gaussian curve enables efficient breakpoints table precomputation, thus the discretization step is trivial in comparison to the single vector lookup operation.

Finally, formal definition of SAX as proposed by Lin et al. [11]: A subsequence C of length n can be represented as a word $\hat{C} = \hat{c}_i, ..., \hat{c}_w$ as follows. Let α_i denote the ith element of the alphabet, i.e., $\alpha_1 = a$ and $\alpha_2 = b$. Then the mapping from a PAA approximation C to a word $\hat{C}$ is obtained as follows:

$$\hat{c}_i = \alpha_j, \quad iif \quad \beta_{j-1} \le \bar{c}_i < \beta_j \qquad (2)$$

2.2   Techniques for Distribution Estimation

In the previous section we discussed internals of the original SAX method. SAX uses a Gaussian distribution to derive the range breakpoints resulting in the generation of an equiprobable set of symbols. As we already mentioned, our aim is to make a new SAX method Gaussian distribution requirement free. In this section we want to mention other methods on how to estimate data distribution and, based on them, improve SAX by a different way of setting the range breakpoints. At the beginning, we want to state a common intuition to graphically represent data distribution - histogram plotting from underlying time series data points.

The histogram is a former nonparametric density estimator with strong use in exploratory data analysis for displaying and summarizing data. Bin width is an important parameter that needs selection prior histogram construction. It is evident that the choice of the bin width has a strong effect on the shape of the resulting histogram. The example for different bin width selection can be found in Fig. 2.

There are several ways to determine optimal bin width $\hat{h}$ with n observed instances such as [22]:

$$\hat{h} = \frac{range\_of\_data}{1 + log_2n} \tag{3}$$

or alternatively more general formula based on Mean Integrated Squared Error (MISE):

$$\hat{h} = \hat{C}n^{-1/3}, \tag{4}$$

where $\hat{C}$ is any selected statistic. Most known example of above mentioned formula is normal reference rule [15, 22]:

$$\hat{h} = 3.49\hat{\sigma}n^{-1/3}, \tag{5}$$

where $\hat{\sigma}$ is an estimate of the standard deviation.

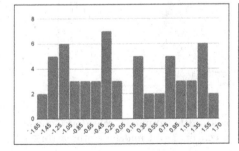 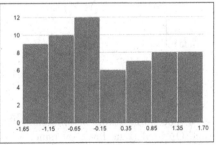

Fig. 2. Comparison of two histograms for the same dataset with bin widths 0.2 and 0.5 respectively. Non-optimally selected bin width causes visually different data distribution.

In the past four decades the research has been performed in the field of distribution estimation using continuous functions, namely density estimators. The intuition behind kernel estimators is to describe underlying data histogram with a smooth continuous line. More formally, given a set of N training data $y_n, n = 1, ..., N$, a kernel density estimator (KDE), with the kernel function K and a bandwidth parameter h, $(h \in R; h > 0)$, gives the estimated density $\hat{f}(y)$ for data y as follows [5]:

$$\hat{f}(y) = \frac{1}{N} \sum_{n=1}^{N} K(\frac{y - y_n}{h}) \tag{6}$$

Kernel function K should satisfy positivity and integrate-to-one constraints [5]:

$$K(u) \geq 0, \quad \int_{R^+} K(u)du = 1 \tag{7}$$

The quality of a kernel estimate depends less on the chosen K than on the bandwidth value h. It is crucial to choose the most suitable bandwidth because a value that is too small or too large will result in not useful estimation. Small values of h lead to undersmoothing estimates while larger h values lead to oversmoothing [6]. Figure 3 illustrates possible cases of incorrectly selected bandwidth parameter h. To overcome this issue, a number of different bandwidth selection methods such as Silverman's rule of thumb [20], Scott's rule of thumb [16] or Improved Sheather-Jones bandwidth selection [18] have been proposed.

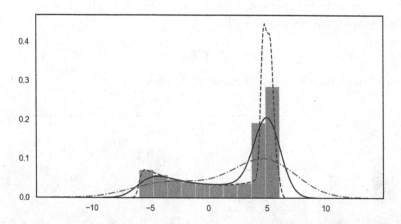

Fig. 3. The implications of different KDE bandwidth parameter selection Gaussian kernel. Optimal (blue line), oversmoothing (green line) and undersmoothing (orange line) bandwidths. (Color figure online)

Table 2 provides a list of most frequently used kernels with their support ranges. Selection of different kernel does not significantly change the shape of the estimated function, but its computational complexity might introduce significant overhead. Support range might be a constraint towards deciding which kernel is suitable for a specific case due to long-tail $\mathbb{R}$ support in cases such as Normal or Laplace kernels.

Table 2. A list of commonly used kernel functions with their function formula and support range [7].

Kernel	$K(u)$	Support
Uniform	$\frac{1}{2\sqrt{3}}$	$\|u\| \leq 3$
Triangular	$\frac{(1-\frac{\|u\|}{\sqrt{6}})}{\sqrt{6}}$	$\|u\| \leq 6$
Epanechnikov	$\frac{3(1-\frac{u^2}{5})}{4\sqrt{5}}$	$\|u\| \leq 5$
Biweight	$\frac{15(1-\frac{u^2}{7})^2}{16\sqrt{7}}$	$\|u\| \leq 7$
Cosine	$\frac{\sqrt{\pi^2-8}cos(\frac{\sqrt{\pi^2-8x}}{2}))}{4}$	$\|u\| \leq \frac{\pi}{\sqrt{\pi^2-8}}$
Normal	$\frac{2\pi}{\sqrt{(2)}}exp(-\frac{1}{2}x^2))$	$\mathbb{R}$
Laplace	$\frac{exp(-\sqrt{2}\|x\|)}{\sqrt{2}}$	$\mathbb{R}$

3 Distribution-Wise SAX (edwSAX)

Formerly, the proposed SAX method is not well suited for time series with non-Gaussian distribution. If we apply Gaussian distribution lookup vector for break-points estimation, we get non-optimal, still feasible, breakpoints. Our modified implementation focuses on the elimination of this deficiency. Algorithm 1 illustrates the overall main flow of the function where the input is sequence for symbolization, word length, alphabet size, bandwidth for kernel estimator and kernel function. Result is SAX representation of input sequence. In the next sections we will discuss specific internals of our method, namely:

1. *probability density function estimation*: given a normalized time series, we need to estimate a probability density function (*pdf*) to proceed with more precise breakpoints;
2. *breakpoints and centroids vector calculation*: having a probability density function, the task is to calculate equiprobable breakpoints covering the domain of *pdf*;

3. *distance measure definition*: to prove our efficiency and indexing capability of our breakpoints vector, we also define a modified distance measure proving lower bound euclidean distance criterion.

Data: Sequence, WordLength, AlphaSize, Bandwidth, KernelFunction
Result: SAX Repr
PAA Sequence = *ApplyPAA*(Sequence, WordLength);
PDF Estimate = *EstimatePDF*(KernelFunction, Bandwidth, Sequence);
Breakpoints Vector = *CalculateBreakpoints*(AlphaSize, PDF Estimate);
SAX Repr = *Map*(PAA Sequence, Breakpoints Vector);

Algorithm 1: edwSAX main algorithm

Our proposed method flow in contrast to SAX expects a training phase (Fig. 4). At this phase, we need to estimate *pdf* in order to correctly calculate breakpoints and centroids vectors. All further steps after training are identical with the former method. Keeping this fact in mind, we can virtually extend any other improved SAX method with our alternative breakpoints vector in order to be efficient on non-Gaussian-like data sets.

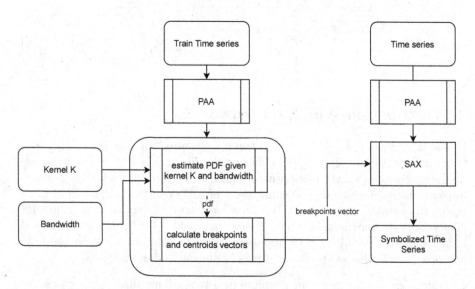

Fig. 4. The flow of edwSAX method. In contrast to SAX, we also need a training phase (left top-down flow) to estimate *pdf* and calculate the breakpoints vector. Once breakpoints vector is trained, we follow well-known SAX flow (right top-down float).

3.1 Probability Density Estimation

With a transformed PAA time series we can proceed to probability density estimation. Having an estimated probability function this will help us further in
the breakpoints vector calculation for the discretization procedure step. A naive
solution to this problem seems to be histogram exploitation as its computational
complexity is incomparably lower to the other methods such as KDE. The main
drawback of histograms is their discrete representation which is not suitable for
breakpoints interpolation.

Our method for breakpoints calculation expects a continuous probability
function suitable for integral calculus with integrate-to-one constraint. KDE
appears to be the solution to this problem. This method needs to specify a kernel function $K(\cdot)$ and a bandwidth parameter h. Selection of appropriate kernel
function and bandwidth parameter depends on data and required precision of
overall symbolic representation performance.

To our best knowledge, Gaussian kernel gives the most relevant results and
should be applied as the first possible option for KDE exploitation. In Table 2 we
list multiple options related to the most frequently used KDE kernels. However,
for our implementation we advise to use Kernels with more compact support
rather than the Gaussian one. The default option is Epanechnikov kernel with
Improved Sheather Jones (ISJ) index which outperforms asymptotically Silverman's rule of thumb on non-Gaussian-like data sets for bandwidth parameter
selection.

The difference between edwSAX and SAX itself depicted in Fig. 5 shows
inefficiently used symbols c, d, e in case of SAX. Symbol e covers whole part of
time series with values alternating around value 1, on the other hand, edwSAX
calculated more efficient symbol breakpoints with alternating symbols d and e.

3.2 Breakpoints and Centroids Vector Calculation

Having estimated probability function using KDE, we can advance and estimate
breakpoints based on probability distribution of time series with their respective representatives - centroids. The main goal is to efficiently compute those
breakpoints as we do not apply only specific Gaussian distribution and its pre-
computed values table. However, the idea for breakpoints selection is the same -
select points from the KDE *pdf* such that they produce equal-sized areas under
the KDE function curve.

Definition 1. *Let a denote alphabet size, pdf probability density function and
β_n, β_{n+1} any two consecutive breakpoints from breakpoints vector B. Then breakpoints vector B is defined as a vector of ordered breakpoints β such that β_n, β_{n+1}
follows:*

$$\int_{\beta_n}^{\beta_{n+1}} pdf(y)dy = \frac{1}{a} \tag{8}$$

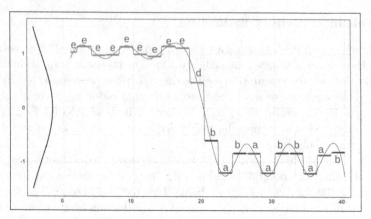

(a) Example of SAX representation.
Resulting symbolic string: *eeeeeeeedbabaabbaab*

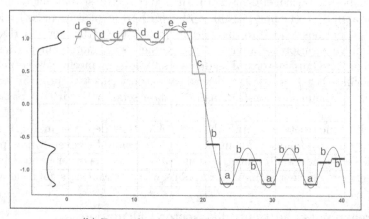

(b) Example of edwSAX representation.
Resulting symbolic string: *deddeddeecbabbabbabb*

Fig. 5. Application of different *pdf* on breakpoints selection. a) SAX with Gaussian distribution, b) edwSAX with KDE (Epanechnikov Kernel ISJ bandwdith). KDE based breakpoints bring more precise symbolic representation compared with SAX with the same alphabet size 5 and PAA = 2. edwSAX covers more efficiently time series values greater than 0 with 3 different symbols (c, d, e) in contrast to 2 SAX symbols (d, e).

With the computed breakpoints vector, we proceed in most suitable representatives - centroids calculation for each consecutive non-overlapping breakpoints pair. Most suitable representative represents in our case centroid γ, $\gamma \in \langle \beta_n; \beta_{n+1} \rangle$, fulfilling equiprobable condition for ranges $\langle \beta_n; \gamma \rangle$ and $\langle \gamma; \beta_{n+1} \rangle$:

$$\int_{\beta_n}^{\gamma} pdf(y)dy = \int_{\gamma}^{\beta_{n+1}} pdf(y)dy \quad iif \quad \gamma \in \langle \beta_n; \beta_{n+1} \rangle \qquad (9)$$

Computed centroids vector is stored and used along breakpoints vector during the reconstruction process, giving us lower reconstruction error by means of any well-known error metrics such as MAE or MSE. Breakpoints vector is also necessary in order to correctly calculate distance measure, defined in Sect. 3.3.

Discretization process follows the same algorithm as proposed in the original SAX method. For reference see Algorithm 2.

Data: PAA Sequence, Breakpoints Vector B
Result: SAX Representation
foreach *Segment in PAA Sequence* **do**
    for $i \leftarrow 2$ **to** *Length(B)* **do**
        if $B[i-1] \leq$ *Segment* **and** *Segment* $< B[i]$ **then**
          | Append(SAX Representation, Alphabet[i])
        end
    end
end

Algorithm 2: edwSAX mapping procedure

3.3 Distance Measure

Having defined the symbolic time series representation, we now define the similarity measure on the transformed data and we prove it is lower bound of the Euclidean distance on the original data. Distance measure for edwSAX time series representation is based on the existing proved MINDIST distance [11] - Euclidean distance adaption:

$$MINDIST(\tilde{Q}, \tilde{C}) \equiv \sqrt{\frac{n}{w}} \sqrt{\sum_{i=1}^{w} (dist(\tilde{q}_i, \tilde{c}_i))^2} \qquad (10)$$

where $\tilde{Q}$, $\tilde{C}$ are symbolized time series, n is original time series length, w is word length (based on PAA parameter) and $dist(\tilde{q}_i, \tilde{c}_i)$ is a function returning minimal distance between symbols $\tilde{q}_i$ and $\tilde{c}_i$.

To make the function MINDIST working, we need to construct a lookup table for Euclidean distance between transformed symbols - an internal *dist* function. From the graphical proof (Fig. 6), it is clear that even in our implementation of $dist(q, c)$ SAX internal lookup $cell(q, c)$ function is applied (Eq. 11). The example of such calculation for the breakpoints vector is listed in Table 3. Zero values on diagonal are due to Euclidean distance nature - distance to itself is always zero. Other special zero distance is between two consequent symbols, depicted in Fig. 6, the only lower bound constrained distance between β_i and β_{i+1} is always zero as upper bound for interval β_i; β_{i+1} is β_{i+1} and lower bound for β_{i+1}; β_{i+2} is β_{i+1}.

$$cell_{q,c} = \begin{cases} 0 & if\, q = c \quad or \quad max(q,c) = min(q,c) + 1 \\ \beta_{max(q,c)-1} - \beta_{min(q,c)} & otherwise \end{cases}$$

$$(11)$$

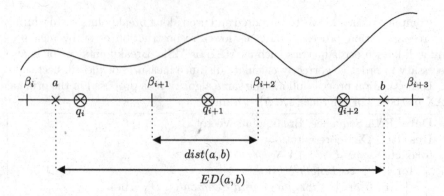

Fig. 6. Graphical proof of mindist function definition. Symbol a might occur anywhere on range $\langle \beta_i; \beta_{i+1} \rangle$ and symbol b on range $\langle \beta_{i+2}; \beta_{i+3} \rangle$. In this case, minimal difference is between lower bound of greater symbol and upper bound of lower symbol I.e $mindist(a, b) = \beta_{i+2} - \beta_{i+1}$. This applies for all cases ranging for symbols located more than one symbol from each other $(\tilde{q}_i; \tilde{q_{i+1}})$ or arbitrary symbols distant $(\tilde{q}_i; \tilde{q_{i+n}})$. The only exception is self-symbol distance $(\tilde{q}_i; \tilde{q}_i)$ and co-located symbols $(\tilde{q}_i; \tilde{q_{i+1}})$ - strictly defined as 0.

4   Experiments and Evaluation

We evaluated our method by means of reconstruction error and tightness of lower bound proof. As far as we know, the most suitable method to compare with is the SAX. In the next sections we describe used evaluation data sets and discuss achieved results with their implications in real life method exploitation.

4.1   Evaluation Data Sets Description

For evaluation purposes, we selected a consistent collection of 20 different time series data sets from UEA & UCR Time Series Classification Repository [1]. Table 4. describes a general overview for all datasets such as their time series

Table 3. The example of lookup table for estimated *pdf*, alphabet size $a = 6$ and breakpoints vector $\beta = [-\infty, -0.33, -0.01, 0.66, 0.97, 1.54, \infty]$.

	a	b	c	d	e	f
a	0.00	0.00	0.32	0.99	1.30	1.87
b	0.00	0.00	0.00	0.67	0.98	1.55
c	0.32	0.00	0.00	0.00	0.31	0.88
d	0.99	0.67	0.00	0.00	0.00	0.57
e	1.30	0.98	0.31	0.00	0.00	0.00
f	1.87	1.55	0.88	0.57	0.00	0.00

count, train/test size, categories count and their type. The selection covers different types of time series such as Image, Sensor, Motion or Simulated; and variable lengths ranging from tens to hundreds of points with overlap into different domains such as biology, physics, chemometrics or industry in general.

Considering our further evaluation, it is necessary to have intuition about underlying data distribution for each evaluation dataset - their shape. Figure 7 shows grid of all estimated data sets *pdf*s using Epanechnikov kernel and per data set calculated ISJ bandwidth parameter. There are several multimodal datasets such as Beef, Coffee or TwoPatterns, several mixed ones with uncertain distribution like GunPoint or Trace, and clear Gaussian-like shaped datasets - BettleFly, OSULeaf or Worms.

4.2   Tightness of Lower Bound

Tightness of Lower Bound (TLB) is a key metric for method indexing performance (Eq. 12). Symbolic or in general dimensionally reduced time series representations are used for cheap database index lookups without any need to access raw timeseries stored on slow large-capacity storage. Optimal TLB ranges values from zero to one; meaning poorly or truly precise calculated symbolized time series distance compared to distance measure calculated on raw time series.

$$TLB = \frac{MINDIST(\tilde{Q}, \tilde{C})}{D(Q, C)} \qquad (12)$$

In our test scenario, we trained for each data set edwSAX model with Epanechnikov kernel, ISJ automatic bandwidth selection and PAA transformation parameter 2. After we trained our models, we evaluate TLB performance using different alphabet sizes such as $[5, 10, 20, 30, 40, 50, 60, 70, 80, 90, 100]$ symbols. For 16 out of 20 datasets we were able to reach pretty similar performance varying TLB from 0.29 to 0.61 for alphabet size 5 to maximum range from 0.79 to 0.98 for alphabet size 100.

From Fig. 8 we can see a steady relative increasing trend between alphabet size and TLB. Reasonable trade-off between alphabet size and gained TLB between 20 and 30 alphabet symbols (Fig. 9). OliveOil as an outlier in performance strongly underperforms this task reaching TLB from 0.03 for 5 symbols to only 0.31 for 100 symbols. TLB Trend is also increasing, but still underperforming the other datasets. If we look at data distribution for this dataset, we can see a strongly fragmented distribution across the whole range of values.

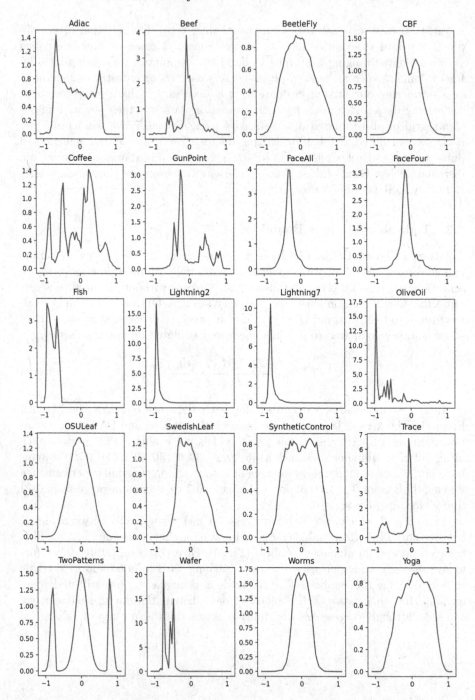

Fig. 7. KDE distributions estimation visualizations using Epanechnikov kernel and ISJ based bandwidth depicting data distributions modality.

Table 4. Brief evaluation datasets description.

Dataset	Train S	Test S	Avg Len	Classes	Type	Domain
Adiac	390	391	176	37	Image	Biology
Beef	30	30	470	5	Spectro	Chemometrics
BeetleFly	20	20	512	2	Image	Biology
CBF	30	900	128	3	Simulated	N/A
Coffee	28	28	286	2	Spetro	Chemometrics
FaceAll	560	1690	131	14	Image	Surveillance
FaceFour	24	88	350	4	Image	Surveillance
Fish	175	175	463	7	Image	Biology
GunPoint	50	150	150	2	Motion	Surveillance
Lightning2	60	61	637	2	Sensor	Physics
Lightning7	70	73	319	7	Sensor	Physics
OSULeaf	200	242	427	6	Image	Biology
OliveOil	30	30	570	4	Spectro	Chemometrics
SwedishLeaf	500	625	128	15	Image	Biology
Syn. Control	300	300	60	6	Simulated	N/A
Trace	100	100	275	4	Sensor	Industry
TwoPatterns	1000	4000	128	4	Simulated	N/A
Wafer	1000	6164	152	2	Sensor	Industry
Worms	181	77	900	5	Motion	Biology
Yoga	300	3000	326	2	Image	Surveillance

We evaluated the relation between alphabet size and TLB precision gained by adding extra symbols to the alphabet. Having average TLB across all data sets for respective alphabet size, we see reasonable improvement of TLB per each extra symbol ranging 0.07 per symbol, alphabet size five, to 0.02 per symbol for alphabet size forty. Using alphabet size over 40 still improves TLB, but this improvement is not significantly comparable to less symbols alphabets.

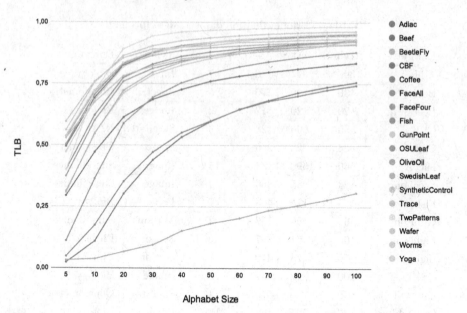

Fig. 8. Tightness of Lower Bound performance with respect to alphabet size ranging from 5 to 100 symbols on all evaluation data sets. With increasing alphabet size, TLB is getting tighter to Euclidean distance.

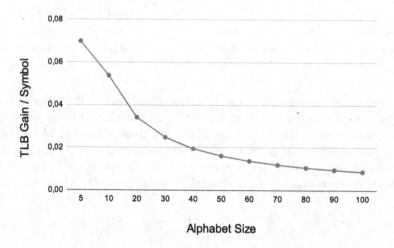

Fig. 9. Ratio between alphabet size and TLB improvement. The more symbols we have in the alphabet, the better the results. However, a reasonable boundary seems around alphabet size 40.

4.3 Reconstruction Error

Given symbolic time series representation, it is essential to evaluate the ability to reconstruct time series back from symbolic representation. Every symbolic representation method introduces various errors by smoothing out time series details/reducing dimensionality. The lesser the reconstruction error the more details encoded in symbolic representation.

In this task, we focused on reconstruction error comparison between edwSAX and SAX methods with the respect to alphabet size evaluating RMSE metric between original and reconstructed time series. Figure 10 show the results comparing SAX and edwSAX for all already mentioned evaluation data sets (Sect. 4.1) for alphabet size 5 and 10 with parameters PAA = 2, Epanechnikov kernel and ISJ bandwidth selection. All mentioned results satisfy statistical significance by Wilcoxon Signed-Rank test for p = 0.05. For alphabet size 5, SAX reached average RMSE reconstruction error 0.43 while edwSAX 0.34. With increasing alphabet size both methods improved their results leading SAX and edwSAX to 0.24 and 0.20 respectively. For detailed results with standard deviations, please refer to Table 5.

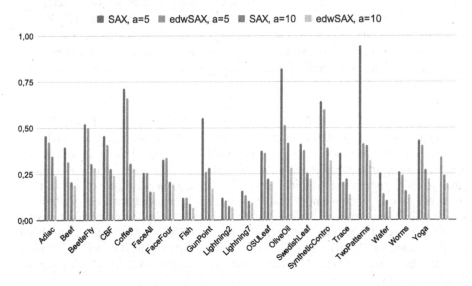

Fig. 10. Reconstruction errors (RMSE) comparison between SAX and edwSAX with respect to alphabet size 5 and 10.

In the second evaluation task we compared reconstruction error with respect to alphabet size. Results (Fig. 11) show that with increasing alphabet size the reconstruction error is decreasing. Significant reconstruction error improvement is up to alphabet size 50, with smaller improvement for larger alphabets.

Table 5. Detailed results for reconstruction error (RMSE) comparison between SAX and edwSAX with respect to alphabet size 5 and 10.

Dataset	SAX, a = 5		edwSAX, a = 5		SAX, a = 10		edwSAX, a = 10	
	AVG	STD	AVG	STD	AVG	STD	AVG	STD
Adiac	0.46798	0.04336	0.43558	0.02608	0.35130	0.03192	0.24581	0.01012
Beef	0.42872	0.05118	0.45018	0.08390	0.22054	0.02548	0.20472	0.03357
CBF	0.44720	0.04410	0.40646	0.04595	0.27328	0.01467	0.24369	0.04257
Coffee	0.72978	0.02463	0.64465	0.02335	0.30757	0.02409	0.26161	0.00927
FaceAll	0.25646	0.00894	0.28358	0.01465	0.15228	0.00907	0.15377	0.00904
FaceFour	0.32128	0.01505	0.34581	0.03310	0.19923	0.02141	0.19627	0.02353
Fish	0.12055	0.00794	0.13240	0.01472	0.09163	0.00576	0.06558	0.00409
GunPoint	0.53585	0.13174	0.27708	0.08271	0.29617	0.14523	0.14883	0.02309
Lightning2	0.11315	0.02612	0.12756	0.03903	0.08079	0.01399	0.08295	0.02091
Lightning7	0.15137	0.02843	0.17274	0.03469	0.11035	0.01227	0.10787	0.02266
OSULeaf	0.38912	0.03333	0.39733	0.03857	0.22264	0.01589	0.21322	0.02027
OliveOil	0.81951	0.00839	0.55603	0.00849	0.41591	0.00354	0.27051	0.00277
SwedishLeaf	0.39812	0.02971	0.37035	0.03210	0.22593	0.01982	0.21155	0.02902
Syn. Control	0.69695	0.05037	0.63973	0.05991	0.38357	0.03309	0.31669	0.02471
Trace	0.37012	0.09105	0.19415	0.08092	0.28719	0.10500	0.12201	0.04162
TwoPatterns	1.03338	0.09201	1.39209	0.19217	0.31282	0.06090	0.30089	0.09275
Wafer	0.23574	0.06626	0.13432	0.05170	0.08131	0.03093	0.07751	0.03280
Yoga	0.47174	0.03554	0.45632	0.05527	0.25553	0.01191	0.23796	0.02052

To sum up achieved results, in both tasks 1) Tightness of Lower Bound and 2) Reconstruction Error edwSAX proved its potential over SAX with significantly better results on data sets that are more varying from Gaussian-like distribution. Our method reaches tightly asymptotically lower bound Euclidean distance with increasing alphabet size. This finding was also proved by average TLB per symbol improvement making our TLB increasing with every addition of symbol. Reconstruction error evaluation also shows improved performance of SAX by means of more efficient symbol representative values selection. Improved symbols selection clearly outperforms SAX on TwoPatterns dataset with strong multimodal distribution having reconstruction error 0.95 while edwSAX achieved error only 0.41.

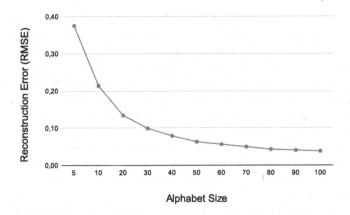

Fig. 11. Reconstruction error (RMSE) trend across all evaluation data sets with respect to alphabet size. (edwSAX setup: Epanechnikov kernel, ISJ bandwidth selection)

5   Conclusion and Future Work

As stated in the Introduction, our main goal was to improve symbolic representation of time series with non-Gaussian data distribution. Lin et al. [11] proposed a superior method for symbolic representation of time series - SAX. Although this approach is interesting, it does not work well for time series with non-Gaussian data distribution. We believe that we have designed an innovative solution for this problem. Our approach extends original SAX by means of dynamically captured data distribution of underlying time series and defining alternative vector of breakpoints for characters mapping. Data distribution estimation at its simplest form could be estimated through well-known and widely applicable histograms. However, this approach suffers from the ease of dynamic computation of breakpoints. An alternative solution, though with high overheads, is estimation using continuous function based estimator. Density estimators appear to be a solution to this problem. The most common variants of these estimators are kernel density estimators.

This method represents a viable alternative to the original SAX method. We empirically proved our method and output symbolic representation with a proper distance lookup table leveraging existing MINDIST distance measure lower bounds Euclidean distance. It makes our method feasible for index lookup operations. We compared our method with original SAX in two tasks: tightness of lower bound and reconstruction error. Both tasks directly/indirectly prove that our method superiors SAX with the most significant difference on multimodal data sets.

The most important limitation lies in unnecessary KDE application in case of highly Gaussian distributed data. Applying both methods in this case will result in very similar symbolic representation. KDE estimates breakpoints similar to pre-computed breakpoints from the SAX table, but with undoubtedly higher computational complexity. On the other hand, applying edwSAX without any prior knowledge of data distribution will safely produce efficient symbolic representation. A number of potential shortcomings need to be considered. Firstly, computational complexity of KDE and breakpoints vector recalculations needs to be considered in case of online exploitation. Secondly, the concept drift at its basis is not covered in the proposed method, though KDE with periodical recalculation is able to overcome a skew in data distribution to some extent. The third shortcoming is connected with the fact that knowledge of breakpoints vector and centroids used during discretization is crucial for further operations such as time series indexing and distance measure in general. These issues are challenges for our future work. Nevertheless, we believe that our work could be a springboard for research in the field of data distribution-aware symbolic time series representation.

This study has gone some way towards enhancing our understanding of efficient symbolic time series representation. To deepen our research we plan to design an online version of our method to tackle computational complexity with hard online processing constraints. Our results are promising and should be validated by a larger sample size time series from real-life environments.

Acknowledgement. This research was supported by TAILOR, a project funded by Horizon 2020 research and innovation programme under GA no 952215 and "Knowledge-based Approach to Intelligent Big Data Analysis" - Slovak Research and Development Agency under the contract No. APVV-16-0213.

References

1. Bagnall, A., Lines, J., Keogh, E.: UEA & UCR time series classification repository (2021). www.timeseriesclassification.com
2. Bountrogiannis, K., Tzagkarakis, G., Tsakalides, P.: Data-driven kernel-based probabilistic SAX for time series dimensionality reduction. In: 2020 28th European Signal Processing Conference (EUSIPCO), pp. 2343–2347. IEEE (2021)
3. Eghan, R.E., Amoako-Yirenkyi, P., Omari-Sasu, A.Y., Frimpong, N.K.: Time-frequency coherence and forecast analysis of selected stock returns in Ghana using Haar wavelet. J. Adv. Math. Comput. Sci. 1–12 (2019)
4. Fournier-Viger, P., Lin, J.C.W., Kiran, R.U., Koh, Y.S., Thomas, R.: A survey of sequential pattern mining. Data Sci. Pattern Recogn. **1**(1), 54–77 (2017)
5. Hwang, J.N., Lay, S.R., Lippman, A.: Nonparametric multivariate density estimation: a comparative study. IEEE Trans. Sig. Process. **42**(10), 2795–2810 (1994)
6. Jones, M.C., Marron, J.S., Sheather, S.J.: A brief survey of bandwidth selection for density estimation. J. Am. Stat. Assoc. **91**(433), 401–407 (1996)
7. Jones, M.C.: The performance of kernel density functions in kernel distribution function estimation. Stat. Prob. Lett. **9**(2), 129–132 (1990)
8. Keogh, E., Chakrabarti, K., Pazzani, M., Mehrotra, S.: Dimensionality reduction for fast similarity search in large time series databases. Knowl. Inf. Syst. **3**(3), 263–286 (2001). https://doi.org/10.1007/PL00011669
9. Keogh, E., Lin, J., Fu, A.: HOT SAX: efficiently finding the most unusual time series subsequence. In: Fifth IEEE International Conference on Data Mining (ICDM 2005), p. 8 IEEE (2005)

10. Kloska, M., Rozinajova, V.: Distribution-wise symbolic aggregate ApproXimation (dwSAX). In: Analide, C., Novais, P., Camacho, D., Yin, H. (eds.) IDEAL 2020. LNCS, vol. 12489, pp. 304–315. Springer, Cham (2020). https://doi.org/10.1007/978-3-030-62362-3_27

11. Lin, J., Keogh, E., Lonardi, S., Chiu, B.: A symbolic representation of time series, with implications for streaming algorithms. In: Proceedings of the 8th ACM SIG-MOD Workshop on Research Issues in Data Mining and Knowledge Discovery, pp. 2–11 (2003)

12. Lin, J., Keogh, E., Wei, L., Lonardi, S.: Experiencing SAX: a novel symbolic representation of time series. Data Min. Knowl. Disc. 15(2), 107–144 (2007). https://doi.org/10.1007/s10618-007-0064-z

13. Mahmoudi, M.R., Heydari, M.H., Roohi, R.: A new method to compare the spectral densities of two independent periodically correlated time series. Math. Comput. Simul. 160, 103–110 (2019)

14. Sato, T., Takano, Y., Miyashiro, R.: Piecewise-linear approximation for feature subset selection in a sequential logit model. J. Oper. Res. Soc. Jpn. 60(1), 1–14 (2017)

15. Scott, D.W.: On optimal and data-based histograms. Biometrika 66(3), 605–610 (1979)

16. Scott, D.W.: Scott's rule. Wiley Interdisc. Rev. Comput. Stat. 2(4), 497–502 (2010)

17. Senin, P., Malinchik, S.: SAX-VSM: interpretable time series classification using SAX and vector space model. In: 2013 IEEE 13th International Conference on Data Mining, pp. 1175–1180. IEEE (2013)

18. Sheather, S.J., Jones, M.C.: A reliable data-based bandwidth selection method for kernel density estimation. J. Roy. Stat. Soc.: Ser. B (Methodol.) 53(3), 683–690 (1991)

19. Shieh, J., Keogh, E.: iSAX: indexing and mining terabyte sized time series. In: Proceedings of the 14th ACM SIGKDD International Conference on Knowledge Discovery and Data Mining, pp. 623–631 (2008)

20. Silverman, B.W.: Density Estimation for Statistics and Data Analysis. Routledge, Boca Raton (2018)

21. Tamura, K., Ichimura, T.: Clustering of time series using hybrid symbolic aggregate approximation. In: 2017 IEEE Symposium Series on Computational Intelligence (SSCI), pp. 1–8. IEEE (2017)

22. Wand, M.: Data-based choice of histogram bin width. Am. Stat. 51(1), 59–64 (1997)

23. Wang, X., Mueen, A., Ding, H., Trajcevski, G., Scheuermann, P., Keogh, E.: Experimental comparison of representation methods and distance measures for time series data. Data Min. Knowl. Disc. 26(2), 275–309 (2013). https://doi.org/10.1007/s10618-012-0250-5

24. Yang, S., Liu, J.: Time-series forecasting based on high-order fuzzy cognitive maps and wavelet transform. IEEE Trans. Fuzzy Syst. 26(6), 3391–3402 (2018)

Anomaly Detection in Time Series

Heraldo Borges(✉), Reza Akbarinia(✉), and Florent Masseglia(✉)

Inria & LIRMM, Univ. Montpellier, Montpellier, France
{heraldo.pimenta-borges-filho,reza.akbarinia,florent.masseglia}@inria.fr

Abstract. Data mining has become an important task for researchers in the past few years, including detecting anomalies that may represent events of interest. The problem of anomaly detection refers to finding samples that do not conform to expected behavior. This paper analyzes recent studies on the detection of anomalies in time series. The goal is to provide an introduction to anomaly detection and a survey of recent research and challenges. The article is divided into three main parts. First, the main concepts are presented. Then, the anomaly detection task is defined. Afterward, the main approaches and strategies to solve the problem are presented.

Keywords: Time series · Anomaly detection

1 Introduction

In many real-world applications including load demand forecasting [25], geography [10], human activity recognition [14], stock return [16] and others [6], the data is collected in the form of time series. Anomaly detection in this type of data refers to discovering any abnormal behavior within the data encountered in a specific time interval. Anomaly detection has been widely used in several application areas. For instance, cardiologists are interested in identifying anomalous parts of ECG signals to diagnose heart disorders [24]. Economists are interested in anomalous parts of share prices to analyze and build economic models[18]. Meteorologists are interested in anomalous parts of weather data to predict future consequences [26].

Consistent with [7], we can define an anomaly as a point in time where the system's behavior is unusual and significantly different from previous normal behavior. An anomaly can mean a change with positive consequences, such as an increase in the number of sales on a sales website, and a change with negative consequences, such as a change in the rotation frequency of a jet engine's turbine. In either case, whether positive or negative, the similarity is that anomalies must often be considered and carefully taken into account.

Anomalies can be temporal if the temporal sequence of data is relevant; i.e., the data maybe anomalous only in a specific temporal context. Temporal anomalies are often subtle and hard to detect in real applications. Detecting temporal anomalies in applications is valuable as they can allow reactions (*e.g.* serve as an early warning for problems with the underlying system, or decision support for adjustments in order to benefit from the anomaly).

© Springer-Verlag GmbH Germany, part of Springer Nature 2021
A. Hameurlain and A. M. Tjoa (Eds.): TLDKS L, LNCS 12930, pp. 46–62, 2021.
https://doi.org/10.1007/978-3-662-64553-6_3

This article aims to provide an organized overview of existing research to detect anomalies in time series. The objective is to provide an understanding of the problem of detecting anomalies and how existing techniques are related. The next section provides some definitions. In Sect. 3, a brief overview of time series anomaly detection techniques is reported.

2   Definitions

A time series is an ordered list of observations $X = < x_1, ..., x_n >$, where each value x_p is an observation collected at time p. Let $S_p = x_p, x_{p+1} + ... + x_{p+m-1}$ be the subsequence of size m, which starts at the position p of the time series X. Figure 1 presents an example of the time series, where $O1$ is an example of a subsequence.

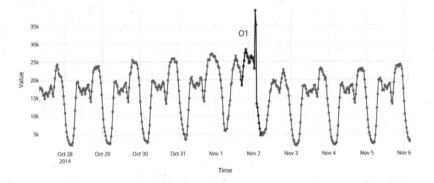

Fig. 1. Time series

2.1   Time-Series Patterns

We briefly describe some main properties of time series important for anomaly detection methods.

We can say that a time series has a **trend** if the observed mean μ is not constant, but varies over time. The trend can have a linear behavior. The time series in Fig. 2 has a positive trend over the years.

The **seasonality** is the recurrence of oscillations periodically. A time series is seasonal if it is influenced by seasonal aspects such as year, period, or year. Thus, there is always a fixed-term, where the oscillation reappears. Figure 3 shows a seasonal time series with annual house sale values. The seasonality can be observed, since the real estate market is generally not active at the beginning of the year and sales generally increase in the middle of the year, decreasing again at the end of the year.

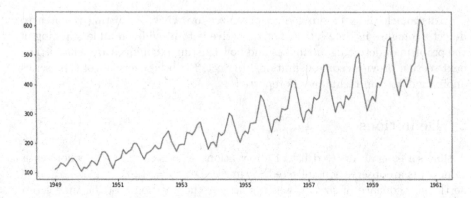

Fig. 2. Sample time-series showing a constant trend

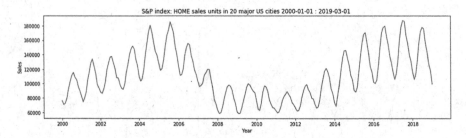

Fig. 3. Sample time-series showing yearly home sales

A **stationary** time series is one whose properties do not depend on the instant at which the series is observed [12]. Thus, time series that have trends, or seasonality, are not stationary, as these properties affect the value of the time series at different times. On the other hand, a series of white noise is stationary because it doesn't matter when it's observed, it looks the same at all times.

The given definition assumes that a stationary series will have the following characteristics:

1. The verification of *constant mean*, shown that there is no trend in the series.
2. Throughout the series period, there is a *constant autocorrelation*
3. The series does not show periodic movements, indicating that there is no *seasonality*

2.2   Anomalies

Several studies have attempted to define the condition of anomalous data. In [11], Hawkins defined anomaly as: "an observation that deviates so much from other observations that it raises the suspicion that a different mechanism generated it." In [23], Barnett and Lewis present another interpretation of this definition: "an outlier is an observation (or subset of observations) that appears to be inconsistent with the rest of this dataset." We can then generalize the definition of anomalies as described in [6] *"patterns in data that do not conform to a well-defined notion of normal behavior"*.

Activities such as a system failure, error in the capture of information by a sensor, or even earthquakes are examples of reasons for anomalies in the data. Even in different scenarios, all of these anomalies produce important information for analysts. The importance of analyzing and understanding these occurrences is an important point for detecting anomalies.

Anomaly Types. We can understand the problem of detecting anomalies for time series to identify outliers concerning some standard or usual signal. While there are many types of anomalies, let's focus on the most important types [6].

1. **Point anomaly**: A point anomaly happens when an observation or a set of several individual observations diverges from the other set of observations. One can observe a punctual discontinuity in a period of time outside the normality of the values. For example, a purchase with a very high amount within a set of credit card transactions. This type is closest to the concept presented in [11].
2. **Contextual anomaly**: An apparently normal observation may diverge when analyzed within a specific context, such as a period of time or a certain location. For example, low temperatures are normal in the winter period but are contextual anomalous when they occur in the summer, where high temperatures are expected.
3. **Collective anomalies**: It occurs when no individual observation is an anomaly in itself, but a group of these observations, when combined, exhibit behavior that diverges from the rest of the data. In the stock market, for example, the fall in the price of an asset does not deviate significantly from the normal range, but the combination of successive falls indicates a collective anomaly.

The contextualization and understanding of the type of anomaly to be identified can help choose the best detection model. Each approach is aimed at certain types of anomalies, having certain advantages and disadvantages according to the subjective definition of normal observations and anomalies in a given context.

Anomaly Detection. Several works provide an overview of techniques for detecting outliers [1]. Anomaly detection is also known as outlier detection, event detection, novelty detection, drift detection, change point detection, failure detection, or misuse detection [6]. These different approaches and terms converge towards the same purpose: detecting abnormal patterns that deviate from the rest of the data, called anomalies or outliers.

Anomaly detection is a broad field that has been studied in the context of a large number of application domains, such as intrusion detection, fraud detection, failure detection in industry 3.0, system health monitoring, event detection in network sensors, and detection of disturbances in the ecosystem [27]. Due to the great diversity of types and techniques, it is necessary to understand the properties of anomalies:

1. Temporality: anomalies can be temporal when associated with some temporal information. These are usually anomalies found in time series as well as streaming data and medical data.
2. Labeling: Some datasets have a set of labeled anomalies, where the location of each instance in the dataset is indicated. In general, this set of labeled anomalies represents only a subset of the anomalies present. Annotated anomalies are used by supervised learning methods, where the method learns through examples to detect new anomalies.
3. Dimensionality: anomalies can present univariate data when represented in only one dimension or multivariate when presenting a set of variables. Multivariate data is generated, for example, on sensors.

Anomaly Detection in Time Series. Detecting anomalies in temporal data differs from detecting non-temporal data. In non-temporal data, as in spatial data, it is possible to detect an anomaly by its location in less dense or more distant regions. Another way is to calculate the deviation from the anomalous observations to the rest of the data. In this domain, it is understood that the observations are independent of each other.

When we look at the temporal data sets, we can see that this premise is not true, as the observations are not completely independent of each other. Previous observations may influence new observations. Thus, the variation between values tends to be smooth and gradual, without major variations.

Take, for example, a list of ten values with the pricing of the value of an asset in the stock market, measured every one minute: $12, $15, $16, $18, $17, $42, $43, $18, $19, $18. Since these points are dependent on each other, we can see that the sudden increase to the value of $42 could be an anomaly.

Anomaly detection methods for time series can be divided into two main categories [1]. There are methods that are based on time series prediction, which represent most of the statistical methods [27]. Another category are methods that are based on unusual forms of the time series. We will present some of these approaches in the next section.

3   Anomaly Detection Approaches for Time Series

In this section, we discuss various anomaly detection applications. We organize the approaches into three categories: statistics-based, cluster-based, and matrix profile-based.

3.1   Statistical Based Approaches

We present some stationary linear autoregressive models. Most of the methods used in time series are linked to linearity and stationarity. A process is stationary when its mean, variance and autocovariance are invariant with respect to time, such as, for example, the Auto-Regressive (AR) and Moving Averages (MA) and Auto-Regressive Moving Averages (ARMA) models.

Autoregressive Model (AR). In an autoregressive model, denoted by AR, we project the variable of interest based on the linear combination of a finite set of values before to the variable (independent variables) and an error value [13]. Thus, an autoregressive model of order p, $AR(p)$, can be written as:

$$Z_t = \varphi_1 Z_{t-1} + \varphi_2 Z_{t-2} + \ldots + \varphi_p Z_{t-p} + a_t \tag{1}$$

where a_t is white noise and the terms $Z_{t-1}, Z_{t-2}, \ldots, Z_{t-p}$ are terms that are independent of a_t. This model assumes that the current value of the series is a linear combination of the past p values of the series and a white noise a_t. We refer to Eq. 1 as an **AR(p) model**, an autoregressive model of order p.

Autoregression-based techniques were extended to detect contextual anomalies in time series. Initially, the AR(p) model is fitted to the data. Then, for each Z_t instance, the residual of the instance is determined for calculating the anomaly score. This residual is the instance value that falls outside the regression model. Thus, the anomaly is scored as the difference between the estimated value and the residual value [6].

Moving Average Model (MA). Instead of using, as in Autoregressive Model (AR), the p observations prior to the variable of interest, a Moving Average Model (MA) considers the last q errors $\epsilon_t, \epsilon_{t-1}, \ldots, \epsilon_{t-q}$ prediction in a regression model. We refer to this model as an **MA (q)**, a moving average model of order q [13]:

$$Z_t = c + \epsilon_t + \theta_1 \epsilon_{t-1} + \theta_2 \epsilon_{t-2} + \cdots + \theta_q \epsilon_{t-q} \tag{2}$$

where the $\theta_1, \theta_2, \ldots, \theta_q$ are the parameters of the model, c is the expectation of Z_t (often assumed to equal 0), and the $\epsilon_t, \epsilon_{t-1}, \ldots, \epsilon_{t-q}$ are the white noise error terms.

It is important to note that this is a moving average model used to predict future values, unlike smoothing models, where average values are used to eliminate some of the randomnesses of the data. Therefore, we consider that each Z_t comes to be considered a weighted moving average of the latest forecast errors.

In the model of Eq. 2, the moving average of the previous q observations, also called the MA of order q, is considered to calculate the current estimated value. The next step includes checking that the estimated value is within the predefined confidence band. Confidence band is the interval defined as a multiple of the standard deviation of the moving average of the previous period. If the value is greater than the maximum expected value (higher confidence band), it is then flagged as an anomaly.

Autoregressive Moving Average Model (ARMA). Autoregressive Moving Average (ARMA) models play a key role in time series modeling, being widely used for time series analysis and prediction. The ARMA model is a combination of the Autoregressive (AR) model, which describes the analytical component of the signal, and the Moving Average (MA) model, which describes the noise component of the signal. Its linear structure also presents a substantial simplification of linear prediction. Compared to pure AR or MA models, ARMA models provide the most efficient linear model in stationary time series, given the ability to model the unknown process with minimal parameterization [28].

An autoregressive and moving average (ARMA) model denoted by ARMA(p, q), combining the AR(p) and MA(q) models, can be written in a single equation in the form:

$$Z_t = c + \epsilon_t + \theta_1 \epsilon_{t-1} + \theta_2 \epsilon_{t-2} + \cdots + \theta_q \epsilon_{t-q} + \varphi_1 Z_{t-1} + \varphi_2 Z_{t-2} + \ldots + \varphi_p Z_{t-p} \quad (3)$$

Let Z_t be the sign Z at time t. We assume that Z_t linearly depends on the previous values $Z_{t-1}, \ldots, Z_{t-p}$ where p is the order of the autoregression. Thus, we have from AR that $\varphi_1, \ldots, \varphi_p$ are auto-regression parameters that can be learned from historical data and used to predict or find similar time series, c is a constant, and ϵ_t $N(0, \sigma^2)$ is Gaussian noise. From MA we have $\theta_1, \ldots, \theta_q$ are learnable parameters of the model, μ is the expectation of Z_t and ϵ_{ti} $N(0, \sigma^2)$ are the terms of Gaussian noise.

Using historical data, we can select p and q in ARMA(p,q) model and learn the model coefficients θ and ϕ based on which we can make a future prediction. A substantial deviation from the prediction result in a time-series anomaly. We can define substantial as two standard deviation from a moving average of Z. The model parameters θ and ϕ are learned via Maximum Likelihood Estimation (MLE). Given the simplicity of the model it may be sufficient for many applications.

The ARMA model has the advantage of having a clear mathematical and statistical basis. However, there are some disadvantage, like the selection of the best values for p and q in order to find a better model for detecting anomalies [4]. If high values are used, the generated model will find a large number of false-negatives, thus, a low number of anomalies. In contrast, using small values for p and q, the generated model will identify a high number of false positives, that is, label a large number of observations as anomalies, which actually are not. There are some ways to adjust the model, such as the use of correlograms, the use of cross-validation [1] and the Box-Jenkins method [3].

3.2   Clustering-Based Approaches

Clustering-based approaches are used to group similar datasets into clusters. Although clustering and anomaly detection tasks are different in nature, several clustering-based anomaly detection techniques have been developed. During the clustering process, these approaches consider the identified observations and clusters and then detect the anomalies [6].

We can group these techniques into two groups based on different assumptions. In first, the dataset is grouped into similar data clusters. Instances that do not belong to any cluster are marked as anomalies. Several algorithms do not force all instances to belong to a cluster; thus, identified clusters are removed from the dataset, and residual instances are noted as anomalies. A disadvantage of these techniques is that they are not optimized to identify anomalies, as their initial objective is to identify clusters. Among these techniques, we have DBSCAN [9] as the most used.

In the second group of techniques, initially, the dataset into clusters, using a clustering algorithm. Then, the distance of each instance to the centroid of the nearest cluster is calculated. Instances that are further away from your centroid are marked as anomalies. A series of anomaly detection techniques that follow this approach have been proposed using different clustering algorithms, among these, the most notorious K-Means clustering [20].

Density-Based Spatial Clustering of Applications with Noise (DBSCAN). Density-based spatial clustering of applications with noise (DBSCAN) is a data clustering non-parametric algorithm [9]. The method is significantly effective in identifying clusters of arbitrary shape and sizes, identifying and separating noise from the data, and detecting *"natural"* clusters and their arrangements within the data space, without any preliminary information about the groups.

Given a set of points in some space, DBSCAN groups points that are intimately very close together (points with many nearby neighbors), marking them as outliers (anomalies) points that are isolated in low-density regions (whose closest neighbors are very far away) [5]. Two important user-defined parameters are required: neighborhood distance epsilon (*eps*) and a minimum number of points *minpts*. For a given point, the points in the *eps* distance are called neighbors of that point. If the total number of points neighboring a point is greater than *minpts*, this group is called a cluster.

DBSCAN labels the data points in three categories: 1) core: the points that have at least *minpts* number of points in the *eps* distance; 2) border: those that are not core points but are the neighbors of core points; 3)outlier (anomalous): those that do not belong to any cluster. Emadi et al. [8] propose an algorithm based on DBSCAN to detect anomalies in Wireless Sensor Networks.

K-Means Clustering. One of the most explored techniques for grouping data is the K-means [20]. It is a cluster based on centroids that partitions the dataset into *k* clusters of similar instances. We can define the k-means algorithm in a sequence of steps. Initially, the number K of clusters is defined. Then, the k centroids are initialized, which can be done by arbitrary choice. Then, for each object, the distance to the centroid of all clusters and connected to the nearest centroid is calculated. Then the centroids of the modified clusters are recalculated. Finally, the distance from the objects to the centroids is recalculated, updating the connections with the nearest ones. This last step is repeated until there are no more updates [17].

Originally, the k-means approach was not defined to work with time series. One of the possibilities to use k-means as anomaly detection in time series is through the use of the sliding window [19] approach. With this approach, a set of subsequences of equal lengths of the same time series is generated. In this set, the k-means algorithm is applied until it converges on the searched k clusters. To perform the anomaly search, the distance between each subsequence and its associated centroid is computed. This subsequence is marked as an anomaly if this distance value is greater than a defined threshold value δ. One of the biggest challenges of this approach is the correct parameterization of the k amounts of clusters.

3.3   Matrix Profile Technique

In 2016, Yeh et al. [29] published a novel technique to perform all-pairs-similarity-search on two time-series, producing two new series: the *Matrix Profile* and the *Matrix Profile Index*. The *Matrix Profile* is defined as a data structure containing the z-normalized Euclidean distances between each subsequence of the first series and its closest corresponding subsequence of the second time series. The Array Profile Index contains the index of the closest matching substring in the second series for each substring. By itself, the Matrix Profile can be used to detect anomalies in contexts where anomalies are defined by unique behavior [22]. In fact, in the Matrix Profile vector the anomalies can be detected in the points with high values, because the distance of the subsequences represented by these points to their closest matching subsequence is high.

In general, given two series of n real values, $S1 \in \mathbb{R}^n$ and $S2 \in \mathbb{R}^n$ and a subsequence length m, the Matrix Profile $M \in \mathbb{R}^{n-m+1}$ and a Matrix Profile Index $I \in \mathbb{R}^{n-m+1}$ are new series such that for each $i \in [0, n-m]$, I_i contains the index of the start of the subsequence of $S2$ with length m that best matches $S1_{i,m}$ and M_i contains the corresponding distance. In the case a *self-join* is performed where $S1 = S2$, an additional constraint is added to prevent *trivial matches*, where subsequences match themselves or nearby subsequences, called *exclusion zone*. The default distance measure used is the z-normalized Euclidean distance, which removes the effect of a changing data offset over time and thus focuses more on shape instead of amplitude. Typical causes of a changing offset are wandering baselines in sensors or natural phenomena (e.g., the gradual change in temperature throughout seasons) [22].

Figure 4 shows an example of the application of the matrix profile in a time series for the detection of anomalies. In this figure, the top image represents a time series, and the bottom image its matrix profile. The top 3 anomalies are marked in red, i.e., the three points where the matrix profile has high values.

Matrix profile is usually able to detect subsequences with unusual shapes in the data. It can be used for detecting point anomalies and also collective anomalies (*i.e.*, a sequence of abnormal points).

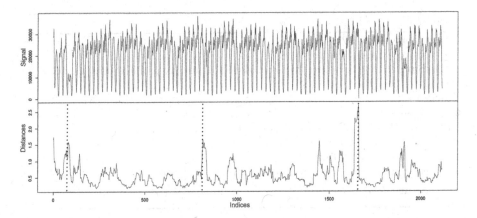

Fig. 4. Matrix profile anomaly detection example

In the rest of this section, we present the main algorithms for calculating the Matrix Profile.

STAMP. The Matrix Profile was originally published together with the STAMP (Scalable Time Series Anytime Matrix Profile) [29], an anytime algorithm to calculate the Matrix Profile over a time series and the corresponding Index. Internally, STAMP uses a similarity search algorithm called MASS [21] that under z-normalized Euclidean iteratively calculates the *distance profile* of each subsequence, which is the distance of the subsequence to every subsequence, by using the Fast Fourier Transform (FFT).

The STAMP is outlined in Algorithm 1. In line 2, the length of T_B is extracted. In line 3, the matrix profile P_{AB} and matrix profile index I_{AB} are initialized. From lines 4 to line 6, the distance profiles D are calculated, using each subsequence $B[idx]$ in the time series T_B and T_A. The pairwise minimum for each element in D is performed with the paired element in P_{AB} (i.e., $\min(D[i], P_{AB}[i])$ for $i = 0$ to length$(D) -1$). Then, as the minimum pair operations are performed, $I_{AB}[i]$ is updated with idx (when $D[i] \leq P_{AB}[i]$). Finally, the results, i.e., P_{AB} and I_{AB}, are returned in line 7. In this format, STAMP computes the matrix profile for the general similarity join. It is possible to change the algorithm to compute the self-similarity join matrix profile of a time series T_A, just by replacing T_B in line 2 with T_A, replace B with A in line 5, and ignore trivial matches in D when performing *ElementWiseMin* in line 6.

The overall complexity of the algorithm is $O(n^2 \log n)$ where n is the length of the time series. Since all subsequences are compared using the MASS algorithm, the $n \log n$ factor comes from the FFT subroutine.

Algorithm 1: STAMP (T_A, T_B, m)

Input: Two time series, T_A and T_B
Interested subsequence length m
Output: A matrix profile P_{AB} and associated matrix profile index I_{AB} of T_A
join T_B

.1 **begin**
2      $n_B \leftarrow Length(T_B)$
3      $P_{AB} \leftarrow$ infs, $I_{AB} \leftarrow$ zeros, idxes $\leftarrow 1 : n_B - m + 1$
4   **for** *each idx in idxes* **do**
5          $D \leftarrow$ MASS$(B[idx], T_A)$
6          $P_{AB}, I_{AB} \leftarrow$ ElementWiseMin(P_{AB}, I_{AB}, D, idx)
7   **return** P_{AB}, I_{AB}

STOMP. The STOMP algorithm is similar to STAMP [15] in that it can be seen as highly optimized nested loop searches, with the repeated calculation of distance profiles as the inner loop. However, while STAMP must evaluate the distance profiles in random order (to allow its anytime behavior), STOMP performs an *ordered* search. It is by exploiting the locality of these searches that STOMP can reduce the time complexity by a factor of $O(\log n)$. STOMP uses the z-normalized Euclidean distance $d_{i,j}$, as shown below, of two time series subsequences $T_{i,m}$ and $T_{j,m}$ using their dot product, $QT_{i,j}$:

$$d_{i,j} = \sqrt{2m \left(1 - \frac{QT_{i,j} - m\mu_i\mu_j}{m\sigma_i\sigma_j} \right)} \tag{4}$$

Here m is the subsequence length, μ_i is the mean of $T_{i,m}$, μ_j is the mean of $T_{(j,m)}$, σ_i is the standard deviation of $T_{i,m}$, and σ_j is the standard deviation of $T_{j,m}$. Note that $QT_{i,j}$ can be decomposed as:

$$QT_{i,j} = \sum_{k=0}^{m-1} T_{i+k}T_{j+k} \tag{5}$$

The time required to compute $d_{i,j}$ depends only on the time required to compute $QT_{i,j}$. To solve this problem, STOMP pre-computes and stores the means and standard deviation in $O(n)$ space and time, thus, it takes $O(1)$ to compute $d_{i,j}$ [30].

The pseudo-code of STOMP algorithm is shown in Algorithm 2. It begins in line 1 by computing the matrix profile length l. In line 2, it calculates the mean and standard deviation of every subsequence in T. Line 3 calculates the first dot

---

Algorithm 2: STOMP (T, m)

Input: A time series T and a subsequence length m
Output: Matrix profile P and the associated matrix profile index I of T

1 **begin**
2      $n \leftarrow \text{Length}(T), l \leftarrow n - m + 1$
3      $\mu, \sigma \leftarrow ComputeMeanStd(T, m)$
4      $QT \leftarrow SlidingDotProduct(T[1 : m], T), QT_{first} \leftarrow QT$
5      $D \leftarrow \text{CalculateDistanceProfile}(QT, \mu, \sigma)$
6      $P \leftarrow D, I \leftarrow \text{ones}$ // initialization
7   **for** $i=2$ **to** l **do**
8   **for** $j=l$ **downto** 2 **do**
9              $QT[j] \leftarrow QT[j-1] - T[j-1] \times T[i-1] + T[j+m-1] \times T[i+m-1]$
10          $QT[1] \leftarrow QT_{first}[i]$
11          $D \leftarrow CalculateDistanceProfile(QT, \mu, \sigma, i)$
12          $P, I \leftarrow ElementWiseMin(P, I, D, i)$
13   **return** P, I

---

product vector QT with the algorithm in TABLE I. Line 5 initializes the matrix profile P and matrix profile index I. The loop in lines 6–13 calculates the distance profile of every subsequence of T in sequential order, with lines 7–9 updating QT according to (5). Then update $QT[1]$ in line 10 is done with the pre-computed QT_{first} in line 3. Line 11 calculates distance profile D according to Equation (4). Finally, line 12 compares every element of P with D: if $D[j] < P[j]$, then $P[j] = D[j]$, $I[j] = i$.

The time complexity of STOMP is $O(n^2)$. Thus, it can achieve a $O(\log n)$ factor speedup over STAMP [15]. The $O(\log n)$ speedup makes little difference for small datasets, however, when considering the datasets with millions of data points, this $O(\log n)$ factor begins to produce a significant performance gain.

SCRIMP++. An extension of *STOMP* is proposed in [31]. *SCRIMP* is an anytime algorithm that computes the matrix profile algorithm combining the anytime component of *STAMP* with the speed of *STOMP*. The optimization of *SCRIMP++* is performed using the incremental calculation of the D diagonals of the scalar product of Eq. 4, as follows:

$$Q_{i,j} = Q_{i-1,j-1} - t_{i-1}t_{j-1} + t_{i+m-1}t_{j+m-1} \tag{6}$$

Equation 6 presents the incremental approach, where the values of the diagonal cells can be calculated using the cell value previously calculated. This approach reduces the number of operations required for the new calculation.

Algorithm 3 shows the pseudo-code of *SCRIMP++*. Line 2 precomputes the means and standard deviations of all subsequences in T. In line 4, matrix profile P and matrix profile index I are initialized. In lines 6–16, the diagonals of the distance matrix are iteratively evaluated, being chosen in random order. Figure 5

Algorithm 3: The $SCRIMP++$ Algorithm

Input: A time series T and a subsequence length m
Output: Matrix profile P and matrix profile index I of T

1 **begin**
2 | $n \leftarrow Length(T)$
3 | $\mu, \sigma \leftarrow ComputeMeanStd(T, m)$
4 | $P \leftarrow inf, I \leftarrow ones$
5 | $Orders \leftarrow RandPerm(m/4 + 1 : n - m + 1)$
6 | **for** k *in* $Orders$ **do**
7 | | **for** $i=1$ *to* $n\text{-}m+2\text{-}k$ **do**
8 | | | **if** $i = 1$ **then**
9 | | | | $q \leftarrow DotProduct(T_{1,m}, T_{k,m})$
10 | | | **else**
11 | | | | $q \leftarrow q - t_1 t_{i+k-2} + t_{i+m-1} t_{i+k+m-2}$
12 | | | $d \leftarrow CalculateDistance(q, \mu_i, \sigma_i, \mu_{i+k-1}, \sigma_{i+k-1})$
13 | | | **if** $d < P_i$ **then**
14 | | | | $P_i \leftarrow d, I_i \leftarrow i + k - 1$
15 | | | **if** $d < P_{i+k-1}$ **then**
16 | | | | $P_{i+k-1} \leftarrow d, I_{i+k-1} \leftarrow i$

17 | returnP, I

---

shows an example of this process. Diagonal distance values, such as $d_{1,k}$, $d_{2,k}$, $\ldots$, $d_{n-m+2-k,n-m+1}$ are calculated one by one. If $d_{i,i+k-1}$, for any $i < n-m+1$, referenced by d in line 12, is less than P_i (line 13) or P_{i+k-1} (line 15), then the associated matrix profile and index values are updated. The iterative algorithm can be interrupted by the user to analyze the matrix profile and index values.

AAMP. In many applications it is preferred to not normalize the time-series data because the anomalies can be detected based on the point values, and not the shapes. AAMP [2] has been designed for such applications. It is an efficient algorithm for computing matrix profile with the pure (non-normalized) Euclidean distance. AAMP is executed in a set of iterations, such that in each iteration the distance of subsequences is computed incrementally. The time complexity of AAMP is $O(n \times (n - m))$ with small constants, where n is the time series length and m the subsequence length. The experiments reported in [2] show that the performance of AAMP is significantly better than that of STAMP and SCRIMP++ (an improved version of STOMP).

The main idea behind AAMP is that for computing the distance between subsequences it uses *diagonal sliding windows*, such that in each sliding window, the Euclidean distance is incrementally computed only between the subsequences that have a precise difference in their *start position*. Let $T_i = \langle t_i, t_{i+1}, \ldots, t_{i+m-1} \rangle$ and $T_j = \langle t_j, t_{j+1}, \ldots, t_{j+m-1} \rangle$ be two subsequences. The sliding windows in AAMP allow to use Eq. 7 for incremental computation of the

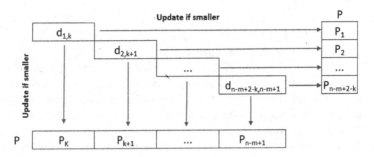

Fig. 5. A SCRIMP++ iteration evaluates a randomly selected diagonal, thus updating the matrix profile

distance between subsequences T_i and T_j (denoted by $D_{i,j}$) by using the yet computed distance between subsequences T_{i-1} and T_{j-1} (denoted as $D_{i-1,j-1}$):

$$[H]D_{i,j} = \sqrt{D_{i-1,j-1}^2 - (t_{i-1} - t_{j-1})^2 + (t_{i+m-1} - t_{j+m-1})^2} \qquad (7)$$

Algorithm 4: AAMP algorithm

Input: T: time series; n: length of time series; m: subsequence length
Output: P: Matrix profile;

1 **begin**
2   **for** $i=1$ to n **do**
3         $P[i] = \infty$;
4   **for** $k=1$ to n-m-1 **do**
5         $dist = Euc\_Distance(T_{1,m}, T_{k,m})$
6   **if** $dist < P[1]$ **then**
7            $P[1] = dist;$
8   **if** $dist < P[k]$ **then**
9            $P[k] = dist;$
10   **for** $i=2$ to n - m + 1 - k **do**
11            $dist = \sqrt{(dist^2 - (t_{i-1} - t_{i-1+k})^2 + (t_{i+m-1} - t_{i+m+k-1})^2}$
12   **if** $dist < P[i]$ **then**
13               $P[i] = dist;$
14   **if** $dist < P[i+k]$ **then**
15               $P[i+k] = dist;$

Algorithm 4 shows the pseudo-code of AAMP. Initially, the algorithm sets all the values of the matrix profile to infinity (*i.e.*, maximum distance). Then, it performs $n - m - 1$ iterations using a variable k ($1 \leq k \leq n - m - 1$). In each iteration k, the algorithm compares each subsequence $T_{i,m}$ with the subsequence that is k positions far from it, *i.e.*, $T_{i,m+k}$. To do this, AAMP firstly computes the Euclidean distance of the first subsequence of the time series, *i.e.*, $T_{1,m}$, with the one that starts at position k, *i.e.*, $T_{k,m}$. This first distance computation is done using the normal formula of Euclidean distance. Then, in a sliding window, the algorithm incrementally computes the distance of other subsequences with the subsequences that are k position far from them, and this is done by using Eq. 7 in $O(1)$. If the computed distance is smaller than the previous minimum distance that is kept in the matrix profile P, then it is updated with this new lower value.

4   Conclusion

In this paper, we presented a survey of anomaly detection methods in time series datasets. We firstly presented the main concepts related to anomalous data in different applications, and then defined the anomaly detection task. Afterwards, we described the important approaches for anomaly detection in three main categories: statistical based, clustering based, and matrix profile based.

References

1. Aggarwal, C.C.: Outlier Analysis. Springer, Cham (2017). https://doi.org/10.1007/978-3-319-47578-3
2. Akbarinia, R., Cloez, B.: Efficient matrix profile computation using different distance functions. CoRR abs/1901.05708 (2019). http://arxiv.org/abs/1901.05708
3. Box, G.: Time Series Analysis: Forecasting and Control. Wiley, Hoboken (2016)
4. Braei, M., Wagner, S.: Anomaly detection in univariate time-series: a survey on the state-of-the-art (2020)
5. Celik, M., Dadaser-Celik, F., Dokuz, A.S.: Anomaly detection in temperature data using DBSCAN algorithm. In: 2011 International Symposium on Innovations in Intelligent Systems and Applications. IEEE, June 2011. https://doi.org/10.1109/inista.2011.5946052
6. Chandola, V., Banerjee, A., Kumar, V.: Anomaly detection: a survey. ACM Comput. Surv. **41** (2009). https://doi.org/10.1145/1541880.1541882
7. Chandola, V., Mithal, V., Kumar, V.: Comparative evaluation of anomaly detection techniques for sequence data. In: 2008 Eighth IEEE International Conference on Data Mining. IEEE, December 2008. https://doi.org/10.1109/icdm.2008.151
8. Emadi, H.S., Mazinani, S.M.: A novel anomaly detection algorithm using DBSCAN and SVM in wireless sensor networks. Wirel. Pers. Commun. **98**(2), 2025–2035 (2017). https://doi.org/10.1007/s11277-017-4961-1
9. Ester, M., Kriegel, H.P., Sander, J., Xu, X.: A density-based algorithm for discovering clusters in large spatial databases with noise. In: Proceedings of the Second International Conference on Knowledge Discovery and Data Mining, KDD 1996, pp. 226–231. AAAI Press (1996)
10. Feyrer, J.: Trade and income—exploiting time series in geography. Am. Econ. J. Appl. Econ. **11**(4), 1–35 (2019). https://doi.org/10.1257/app.20170616

11. Hawkins, D.M.: Identification of Outliers. Springer, Netherlands (1980). https://doi.org/10.1007/978-94-015-3994-4
12. Hyndman, R.: Forecasting: Principles and Practice. OTexts, Heathmont (2014)
13. Hyndman, R.: Forecasting: Principles and Practice. OTexts, Melbourne (2018)
14. Ignatov, A.: Real-time human activity recognition from accelerometer data using convolutional neural networks. Appl. Soft Comput. **62**, 915–922 (2018). https://doi.org/10.1016/j.asoc.2017.09.027
15. Keogh, E., Kasetty, S.: Data mining and knowledge discovery **7**(4), 349–371 (2003). https://doi.org/10.1023/a:1024988512476
16. Koijen, R.S., Lustig, H., Nieuwerburgh, S.V.: The cross-section and time series of stock and bond returns. J. Monetary Econ. **88**, 50–69 (2017). https://doi.org/10.1016/j.jmoneco.2017.05.006
17. Kumari, R., Sheetanshu, Singh, M.K., Jha, R., Singh, N.: Anomaly detection in network traffic using k-mean clustering. In: 2016 3rd International Conference on Recent Advances in Information Technology (RAIT). IEEE, March 2016. https://doi.org/10.1109/rait.2016.7507933
18. Lahmiri, S.: A variational mode decompoisition approach for analysis and forecasting of economic and financial time series. Expert Syst. Appl. **55**, 268–273 (2016). https://doi.org/10.1016/j.eswa.2016.02.025
19. Li, J., Izakian, H., Pedrycz, W., Jamal, I.: Clustering-based anomaly detection in multivariate time series data. Appl. Soft Comput. **100**, 106919 (2021). https://doi.org/10.1016/j.asoc.2020.106919
20. Likas, A., Vlassis, N., Verbeek, J.J.: The global k-means clustering algorithm. Pattern Recogn. **36**(2), 451–461 (2003). https://doi.org/10.1016/s0031-3203(02)00060-2
21. Mueen, A., et al.: The fastest similarity search algorithm for time series subsequences under Euclidean distance, August 2017. http://www.cs.unm.edu/~mueen/FastestSimilaritySearch.html
22. Paepe, D.D., et al.: A generalized matrix profile framework with support for contextual series analysis. Eng. Appl. Artif. Intell. **90**, 103487 (2020). https://doi.org/10.1016/j.engappai.2020.103487
23. Pincus, R.: Barnett, V., Lewis T.: Outliers in Statistical Data. 3rd edn. Wiley (1994). XVII. 582 pp., £49.95. Biometrical Journal 37(2), 256–256 (1995). https://doi.org/10.1002/bimj.4710370219
24. Pourbabaee, B., Roshtkhari, M.J., Khorasani, K.: Deep convolutional neural networks and learning ECG features for screening paroxysmal atrial fibrillation patients. IEEE Trans. Syst. Man Cybern. Syst. **48**(12), 2095–2104 (2018). https://doi.org/10.1109/tsmc.2017.2705582
25. Qiu, X., Ren, Y., Suganthan, P.N., Amaratunga, G.A.: Empirical mode decomposition based ensemble deep learning for load demand time series forecasting. Appl. Soft Comput. **54**, 246–255 (2017). https://doi.org/10.1016/j.asoc.2017.01.015
26. Soares, E., Costa, P., Costa, B., Leite, D.: Ensemble of evolving data clouds and fuzzy models for weather time series prediction. Appl. Soft Comput. **64**, 445–453 (2018). https://doi.org/10.1016/j.asoc.2017.12.032
27. Thudumu, S., Branch, P., Jin, J., Singh, J.: A comprehensive survey of anomaly detection techniques for high dimensional big data. J. Big Data **7**(1) (2020). https://doi.org/10.1186/s40537-020-00320-x
28. Wang, X., Ahn, S.H.: Real-time prediction and anomaly detection of electrical load in a residential community. Appl. Energy **259**, 114145 (2020). https://doi.org/10.1016/j.apenergy.2019.114145

29. Yeh, C., et al.: Matrix profile I: all pairs similarity joins for time series: a unifying view that includes motifs, discords and shapelets. In: 2016 IEEE 16th International Conference on Data Mining (ICDM), Los Alamitos, CA, USA, pp. 1317–1322. IEEE Computer Society, December 2016. https://doi.org/10.1109/ICDM.2016.0179. https://doi.ieeecomputersociety.org/10.1109/ICDM.2016.0179

30. Zhu, Y., et al.: Matrix profile II: Exploiting a novel algorithm and GPUs to break the one hundred million barrier for time series motifs and joins. In: 2016 IEEE 16th International Conference on Data Mining (ICDM), pp. 739–748 (2016). https://doi.org/10.1109/ICDM.2016.0085

31. Zhu, Y., Yeh, C.C.M., Zimmerman, Z., Kamgar, K., Keogh, E.: Matrix profile XI: SCRIMP++: time series motif discovery at interactive speeds. In: 2018 IEEE International Conference on Data Mining (ICDM), pp. 837–846 (2018). https://doi.org/10.1109/ICDM.2018.00099

Designing Intelligent Marine Framework Based on Complex Adaptive System Principles

Amjad Rattrout[1]([⊠]) [iD], Rashid Jayousi[2]([⊠]) [iD], Karim Benouaret[3] [iD],
Anis Tissaoui[4], and Djamal Benslimane[3]

[1] Arab American University, Jenin, Palestine
amjad.rattrout@aaup.edu
[2] Al-Quds University, Jerusalem, Palestine
rjayousi@staff.alquds.edu
[3] Université Claude Bernard Lyon 1, Villeurbanne, France
{Karim.Benouaret,Djamal.Benslimane}@univ-lyon1.fr
[4] Université de Jendouba, Jendouba, Tunisia
anis.tissaoui@fsjegj.rnu.tn

Abstract. The increasing dependency in the maritime transportation platforms brings into the spot the reliability aspect in container terminals (CT). In this paper agent's technology integrated with decision support process in CT complex system. It is used to build intelligence in the containers as to manage the risk at its different levels. In this paper we address the containers control as a system where decision making is distributed among the various components especially the containers carrying dangerous materials. The multi-agent modeling is well suited to represent the complexity as a distributed system. The proposed model based on a complex process which targeting the highly risked containers based on multi agents' systems and intelligent environment called CTRMS: Container Terminal Risk Management System. Diversity of communities and emergence play a major role in this approach. In this paper we define the CT environment as complex adaptive system (CAS) with respect to some CAS properties and mechanisms as aggregation and tagging in CT. The originality of our approach lies in the proposed new mechanisms for reasoning and coordination of agents in which we put special emphasis on the mechanisms of decision making for dynamic and autonomous system components.

Keywords: Complex adaptive systems · Multi agent systems · Maritime systems · Risk management

1 Introduction

The increasing dependency in the maritime transportation platforms brings into the spot the reliability aspect in container terminal (CT). In this paper a new

© Springer-Verlag GmbH Germany, part of Springer Nature 2021
A. Hameurlain and A. M. Tjoa (Eds.): TLDKS L, LNCS 12930, pp. 63–76, 2021.
https://doi.org/10.1007/978-3-662-64553-6_4

method for building intelligence in the containers to manage the risk in its different levels using agent's technology and integration of decision support process in CT complex adaptive system (CAS) is presented. We addressed the containers control system as adaptive one where decision making process was distributed among the various components especially the containers carrying dangerous materials. The multi-agent systems (MAS) modeling is well suited to represent the complexity of distributed systems. We propose a model based on process for targeting the highly risked containers based on multi agents' systems and in intelligent environment called Container Terminal Risk Management System (CTRMS). Diversity of communities and emergence play a major role in this approach. We define the CT environment as CAS with respect to some CAS properties and mechanisms as aggregation and tagging in CT. Any product consists of three separate dimensions (core product, actual product, extended product). The combination between them forms the final product existed and represented in the container. Those presented dimensions should be radically considered at any product definition. In some cases, container may not explicitly mention the physical products name, that could reflects the benefit of the product to customer. The actual product is the real physical product, and finally the extended product, or the non-physical part of the product, that contains information details represents the actual product.

Regarding last definitions and with respect to the distributed complex architecture of product systems, product information management and life cycle management implies on heterogeneous systems with potential conflict formats and in different parts. During our work as researchers in different maritime projects, such as CTRMS which is based on expert systems and lacks independent decision model, therefore we thought a new intelligent framework is needed. In this paper we propose such an intelligent framework based on Multi Agent System (MAS) concepts. In the next section we discuss related work, while in Sect. 3, we give an overview of the proposed framework, in Sect. 4 we present emergence property within the complex system, while in Sect. 5 we discuss 5 the risk in intelligent environment, in conclude with the last section by pointing out the advantages of using MAS for such domain.

2 Related Work

Many definitions of Intelligent Products (IP) are proposed in different ways, but in general all these definitions were presented while we designed the intelligent product (IP) in the physical and information layers. Therefore, the physical product represented in the product itself and the information stored in the database, as well as the intelligence existing in the decision making level [13]. In general, the entire intelligent product was defined with the following properties: Unique Identifier, capable communicating with its own environment, can store it's own data, deploying language to represent features and production requirements, and finally can participate or make decisions relevant to its own density. Two intelligence levels were proposed by Wong, the first level based on the first three properties where the second level covers all the properties mentioned before. Concerning McFarlain [12], a problem rises when he represents a product with no physical entities or virtual products

[11] uses the life cycle as the keyword, where he represents the IP with ID unique, connects to information resources, using lookup mechanisms, and adds to the last a proactively property. As well, he represents the intelligence in two levels as Wong with narrow differences. In the last few years, the emerging of using the term "Intelligent Product" in industrial systems, to capsulised a product with integrated technology that enables it to develop certain capacities in dynamic environment. An IP possesses could characterised the following features: (1) Unique product identity; (2) Specific communication with its static or dynamic environment; (3) Specific capacity to register its own data; (4) Using a digital language to display its features, requirements, etc.; (5) The capacity of participating in making decisions regarding the process.

Furthermore, by applying these characteristics, we could define any product intelligence within the following two levels: The information-oriented intelligence level and decision-oriented level. The first one enables the product to be aware about its status and communication, where intelligence is influenced by the physical data related to the product such as the ID, the communication and finally the capacity in registering its own information in the database. While in the second level, the product can influence and control operations related to its chain production process, where decisions are oriented. In order the product achieve this level of intelligence the product should ensure all features of the list above.

Integrating the Intelligent product to the risk management CT system was a step to create a self organized and intelligent system in the seaport environment in order to come over the terrorist action or illegal transportation for dangerous materials. The involvement of a numerous international actors in the transport of containers makes it vulnerable to terrorism and theft acts. For this purpose, Chatterejee [3] focused on the terrorist acts involving marine containers and investigated possible preventive measures, such as container inspections, to mitigate possible risks. [2] has proposed a study to define the criteria to select vessels that should be inspected by port authorities based on the results of the previous inspections. The study carried out by [7] proves that the majority of risk causes are assigned to packaging deficiencies and related to consolidation activities. Risk management in CT helps to improve the liability of the global transport network. Great attention has been concentrated on the prevention of possible accidents during handling operations in CT, the identification of risk sources and their consequences. [18] had investigated the assessment of the consequences of hazardous materials accident scenario in seaport, such as toxic gas dispersion and fireball events, using simulation models. Furthermore, the study carried out by [14] aims to simulate the occurrence of a terrorist attack in the hazardous materials transport in urban areas, using a dynamic geo-event approach. The construction of a risk scenario and the estimation of its frequency and consequences are the basis of the risk management. Thus, [19] had tackled the estimation of events frequency in an accident scenarios based on the analysis of the historical accident scenarios in seaport and the use of event tree.

[8] had proposed a statistical study about the impact of the human factor on the CT accidents and how the experience of workers may reduce the occurrence of accidents. The research work carried out by [4] shows that the evolution of the number of accidents in port is related to the evolution of container traffic. Moreover, they present that the accident occurrence is more concentrated in the loading and unloading operations of containers, including the stacking operations in storage area. In Business Process Modeling (BPM), [5] defines the business process as a sequence of activities that aim to propose a specific solution for one customer. Object Management Group has proposed the Business process Modeling Notation (BPMN) as schematic standard to facilitate the business process comprehension. Some researchers, proposed architectures for a simulation models adopting a Multi-Agent Systems (MAS) paradigm to build the entities in a adaptive domain. Others proposed solutions solving scheduling and control problems that are related to shipping, stacking containers and controlling the CTs using agent technology [20]. [10] proved that the more the cooperation between shipper agents and CT operator agents is high, the faster the handling of containers will be.

3   Framework Structure

Intelligent activities are complex and abstract by nature. The idea is to clarify what intelligence produces and how to build intelligence in products? The procedures of building intelligence in product should be defined as sub process in the whole of the intelligent system. Therefore, we need to explain in explicit way the dimensions that intelligent products have. We need also to know how to characterize intelligent products using the existing resources to build intelligent resources. To achieve intelligent product we need intelligent process based on continuous process at producing actionable information in order to make well decisions. This cycle of process covers information needs, and the way we collect this information (intelligent collect orientation). The level of trust depends on the level of analyzing and dissemination of information, and how we use the outputs. The dimension vary depends on users and usage purposes of intelligent product. The dimension can be covered: information content, lifecycle, and degree of analysis and frequency. The study of current seaport systems shows the need to develop new models for dynamic self-organized management system particularly in container terminal areas for better use of existing resources. The approach we recommend is based on new models of distributed reasoning, dynamic and cooperative. As part of this research, we propose to address the containers control as a system where decision making is distributed among the various components especially the containers carrying dangerous materials. The multi-agent modeling is well suited to represent the complexity of distributed systems (Fig. 1).

The originality of our approach lies in the proposed new mechanisms for reasoning and coordination of agents in which we will put special emphasis on the mechanisms of decision making for dynamic and autonomous system components. These models take into account the characteristics inherent in their type, mobility, location, etc. Such a need can be addressed by adopting the MAS approach as it involves the coordination of multiple interacting agents that are flexible, robust and scalable that are to complete the required tasks. With flexibility we can generate effective solutions to different problems. Usually MAS systems are robust as in the case of failures other agents can compensate for such damage. Such a need can be addressed by adopting the MAS approach as it involves the coordination of multiple interacting agents that are flexible, robust and scalable in order to complete a single task. With flexibility we can generate effective solutions to different problems. Usually MAS systems are robust as in the case of failures other agents can compensate for such damage. Scalability can be achieved as new agents can be added to the system and working properly with these new elements. This can be provided in a single agent system by changing the elements of the agent but this is not a good solution because this solution has many disadvantages such as cost, re-design and integration. On the other hand, this process can be handled by just adding a new agent to the system in a multi agent system.

As part of this work, we place ourselves in the context of decision support systems for the anticipation (prejudices or information) about the events leading to risks. We will consider the conflicting interests and security depending on the type of products, etc., inherent in the application context of containers import export process. The data for each entity represented in the system, namely intelligent product. To do so, coordination mechanisms by negotiation or by forming coalitions between agents of these competitive systems represent a major challenge for this research and will allow us to achieve our goals. The agent paradigm maps neatly onto many modeling scenarios because each individual in the scenario can be directly represented as an agent in the environment.

Behaviors can be modeled into the agents so that individuals behave in the same way as the entity they are modeling and developing. An interesting point is that individuals are given simple behaviors, and when many agents are simulated as a group, behaviors often emerge that were not explicitly developed into the agents; these are known as emergent phenomenon. Multi-agent simulation can allow parameters of the individuals and the environment to be varied easily, therefore it is possible to experiment with many alternatives and have results in real-time. An intelligence product is difficult to define because it actually is both a product and a multi layer process, even; it is complex and multidimensional by nature. The product is defined in general as a physical product contains some information, where is the actual information outcome of the process. The process, on the other hand, is a systematic way of producing this outcome. The intelligent product is not also a computer software product or a database even if we can in somehow define the database as intelligent product. The cycle for the process should distinguish our real needs, how we gather the information, analysis and information dissemination, and finally how we using it (Fig. 1). The degree of intelligence depends on the degree of information analyzing, the highest is the level the highest is the intelligence (Fig. 3). The products at the law levels provide data without analysis.

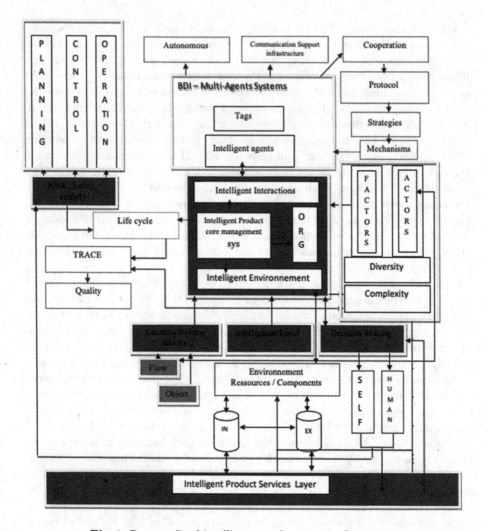

Fig. 1. Decentralized intelligent product core architecture

This kind of information is provided from external or internal resources. Gathered information then analyzed and verified, they can be called analysis-oriented intelligence products. This information should be gathered in high updating frequency, otherwise will not be considered as real time intelligent product process.

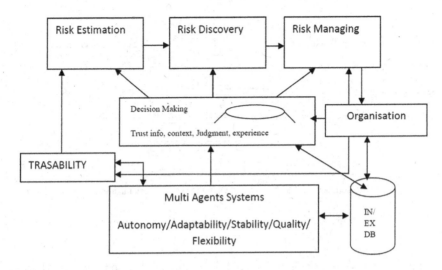

Fig. 2. Risk levels through decision making and MAS

4 Emergence in Complex System

In term of emergence, we aim for positive emergence, which means a subsystem interaction that leads to a better achievement of objectives of the total system than it is explicable by considering the behaviour of every single CT system element. In the context of autonomous logistics, these effects are incorporated by implementing logistic objects (e.g., means of transport, freight, parts) as decentralized subsystems that dynamically coordinate with other subsystems to manage logistic processes and reach their respective goals (e.g., on-time delivery or minimisation of delivery times). Using MAS to implement and model actual systems make it possible to handle, to observe and to improve understanding of complex phenomena. The most of the modern systems requires complexities and high degrees of smoothness which can be achieved using MAS and thus, the system design and implementation become more and more performing and flexible.

Seaports are evolving in many ways and this evolution comes together with extreme diversity. From evolution to diversity and vice versa, we can observe a kind of emergence that occurs in the seaports creating rich and dynamic environment. In this respect, we want to analyze the IP environment, IP interaction and IP organisation as whole as intelligent core with respect to CAS as defined by J. Holland in the model (ECOH) presented in [16].

IP has major characteristics, and in order to better understand these characteristics, we think that it should be perceived as a Complex Adaptive System (CAS). For this reason, an interesting way to address the problem of the complexity in the CIPS, is to create a virtual environment, as a reflection of the real existing one, and where the IPS component is seen on one hand from the view of dependency and on the other hand from the view of combination. In addition,

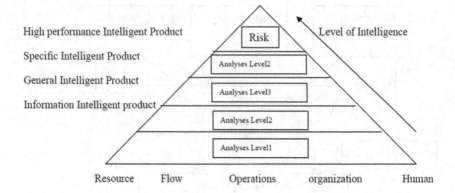

Fig. 3. Managing risk in intelligent environment

we think that the evolution of using the IP is strongly linked to human needs. This evolution should be quantified with the variation of mass information distribution and the consumers. The interàction between these information and consumers is bi-directional and influents the models structures. The network flow is a complex dynamic network, presenting in its properties a self-organizing adaptive behaviour similar to a complex system [15, 17].

The system is going to be modeled in a bottom up approach, and taking these basics into consideration.

4.1   Containers Tagging and Aggregation

What is meant by Aggregation is the Information and results given by the system coming from different resources, actors, and inter-relational objects and populated into a structured of seaport CT database, these actors and objects are forming a complex system community specially in container terminal. While Tagging is the mechanism in which each information unit has a tag or attribute used for identifying and making it easy to construct aggregates through utilizing the used attributes. This tag could tags also type of transporter, type of crain, a human beings (drivers or path, ...). The decision process quality for targeting suspect containers based on the availability of information describing the container's history. The proposed model was enriched by the integration of data collection phase. In order to adapt the decision process according to each set of containers, we aggregate containers with the same description. To achieve this goal, we use the tagging mechanism in order to attribute a specific tag to the containers having the same description. The aggregation step of containers is structured in two phases: The first phase is ensured by two types of agents "Agent On" and "Agent Off". The set of "Agent On" ensures the identification of containers tracked by The GOST system. These containers are equipped with a GPS box and embedded sensors. The set of "Agent Off" provides information about foreign containers. The collected information describes the history of containers on the containers arrival upstream to CT. This step, our agents aggregate containers according to their information (complete and non-complete). The second phase is ensured by the custom agent who tags abnormal containers, unsealed containers.. etc., during the preliminary inspection by the custom officers. Thus, a new set is generated. The container tagging stage aggregates three types of containers; sealed containers with complete description, sealed containers with incomplete description and unsealed containers. Therefore, these results are used by supervising subsystem to take a decision.

The system behaviour is a result of a non-proportionate response to its stimulus. The traffic in seaport charge in natural non-linear, Trucks, Tains and Ships can join the seaport in non linear way, modification from one state to another is non-linear, and so as in the growth of the charging/discharging in the stacking zone, the growth of the system is a nonlinear process too. Flows are the physical resources or the information circulating through the nodes of a complex information network. This can be seen in user's skills, experiments, and backgrounds; also, in system working environments, as traffic-system, this can reflect the diversity of the flow in general and ensure the dynamic adaptive behaviour of a complex adaptive system. Diversity could be observed in the different types of resources. Internal models or schemas are the functions or rules the users can use to interact with themselves and the environment. The component parts which include the building blocks that can be combined and reused for each instance of a model. Identifying these blocks is the first step in modelling a CAS The operations carried out in the seaport terminal are included in the most complex tasks of the transport network systems. This complexity refers to: • The great diversity of entities acting in the container import and export processes. • Interaction and communication with a dynamic environment. • The distributed environment which is formed by a set of independent and interdependence between the different agents and their environment, but whose individual decisions also directly affect the performance of the others. Therefore, it is very difficult to analyze and to develop a single application that holds all the functions. For that we deal with each task independently without losing sight of the close relationship among the tasks.

5 Risks in Intelligent Environment

Our approach aims to tackle the complexity of risk management process in the CT and to ease the analysis of the CT function that exhibit as a CAS. To this end, we use MAS to represent the CT's major components and actors.

Creating a risk management model based on Multi agents system and decision making system would enrich the process of creating an intelligent product system; more the intelligence level is increased more the risk measuring is involved.

The main steps on this structure should take in consideration three items; the risk estimation, risk discovery, and finally the risk management. To estimate a risk we need to have mining tools to estimate the risk in progress. These tools are based on the Container Export Process (CEP) modelling and analyzing system related to. Next, we analyze the risks emerging to the CEP.

CT is prone to a myriad of risks arising from both external and internal causes. Thus, the risk scenario causes identification and its possible consequences are the basis for intervention actions prioritizing. For this purpose, we build credible risk scenario based on the approach illustrated in Fig. 3.

The combination of the events allows us to constructs a set of possible steps. The case of criminal attacks or accidents on the CT is one of the most important scenarios because it leads to the suspension of activities and consequently it causes a big financial loss. For this purpose, we aim to mitigate the occurrence of this risk using a preventive solution based on the integration of CT control barriers. Therefore, we propose to specify a decision support system that assists the custom officers to analyze declarations to target suspect containers.

The next step aims to specify countermeasures to prevent the occurrence of the identified risk scenarios. Finally we deal with the integration of the risk preventive actions in the CT operations in order to assess the impact of risk prevention on the CT performance. All these steps presented in our system agent paradigm. Figure 1 illustrates an overview of our approach.

The functional analysis business process specification represents the basis of this study because it allows us to understand the functioning of the CT and to identify its actors. According to Barjis [1,6] business process modeling will become even more crucial as systems grow in scale and complexity.

Thus, to overcome the complexity of this task and to achieve an accurate description of the studied system, this analysis phase is structured into two main steps; the first one is based on identifying the principal actors and factors functions. The second one aims to model the sequence of tasks composing a business process using the BPMN [1,9,15,17].

In order to identify risk events, we analyze the functioning of CT system and the interaction of its components. Thus, the business process models provided by the previous phase are used to specify the potential risks that threat the CT. Moreover, the identification of the causes and consequences of the identified risk events represents the basis for the specification of possible risk scenarios. Therefore, to achieve an accurate risk analysis and to build realistic risk scenarios we propose to use of the bow-tie approach used in.

We propose the adoption of a decision process based on rules. These rules represent the knowledge acquired by custom officers during their previous interventions. In addition, to go beyond the simple application of an expert system and to ensure the evolution of the decision process, we enrich our system by the integration of an association rule mining method. This method is applied to extract association rules from massive primary data.

Finally, Risk management in our approach helps to improve its reliability and to attract new container flows. However, the application of these measures may slow down the handling operations and consequently affect the CT performance. Thus, in order to assess this impact we propose the integration of the decision process for targeting suspect containers in the functioning of the

CTRMS. In reality, suspect containers targeting by the custom officers is based on the analysis of numerous attributes describing the cargo. Furthermore, the customs officer experience and their ability to analyze numerous attributes are the key factors for the success of decision process. Thus, the experience of the customs officer is represented by the association rules stored in the rule base [15,17].

6 Discussion and Conclusion

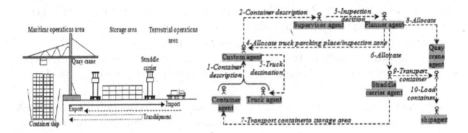

Fig. 4. An overview of the handling process in the CT

Our approach aims to tackle the risk management process in the CT and to ease the analysis of the CT functioning that exhibit as a CAS. To this end, we use MAS as illustrated in Fig. 4 to represent the CT's major components and actors [15,17].

Technologies related to multi agent systems, make it possible for the involvement of the product in decision making protocols at the law level. Then the product becomes an active entity in the decision making process, being able to take control of its own destiny. In our model we considered the Intelligent Product as the aggregation that combines both the material part of a given product and the set of information related to the product. The main concept of the intelligent product: what one should expect of a given product and how we use it. In our model, IP construct is used to represent the concept of the product (the core product) wherever a product illustration is needed.

The term intelligent products stand in these days for the future quality of new products that have some properties as autonomous products. In this work, we are trying to focus on the intelligent product, more precisely, the risk and it's affection on modelling intelligent products systems, taking on considerations these systems as complex environments. To achieve this goal, we are going to study the subject from Risk management point of view using multi agents systems as tools to understand the objects behaviours that composite the modelling process. Therefore, we propose a model that combines multi agents systems (MAS) and complex systems in order to design models that are sufficiently simple and that the mechanisms of emergence can be understood and yet elaborate enough to show interesting behaviour. To design our complex space architecture, the system has been divided into two spaces according to their main tasks. Therefore, many kind of agent for each one of the main job has been modeled. These agents are mainly characterized by their independence from the rest of the system elements. They can coordinate and communicate some simple decisions to the rest of the system. The communication between agents is done by means of asynchronous messages, which are based on FIPA-ACL standard. This design, which is designed according to the agent paradigm, allows us to divide the problem into sub problems. Each sub problem can be solved by a specific agent. In this way, the design is simplified and a highly robust development is allowed. In this case, each agent can be considered as autonomous reasoned. The multi-agent system model used to deal with the design and development of an application which is open, adaptable to the environment, dynamic and robust enough for the efficient risk management of a container terminal.

In our approach we have two types of agents in our environment: • Reactive agent that answers to its environment. • Proactive agent that has to be able to try to fulfill his instructions.

The operations carried out in the seaport terminal are included in the most complex tasks of the transport network systems. This complexity refers to the great diversity of entities acting in the container's import and export processes Interaction and communication with a dynamic environment and the complex environment formed by sets of independent and interdependence agents. Hence, individual decisions also directly affect the performance of the others. To design our agent space architecture, the system has been divided into three spaces according to their main tasks. Therefore, a unique agent created for each task. These agents are mainly characterized by their independence from the rest of the system elements. They can coordinate and communicate some simple decisions to the rest of the system. Risk assessment approach seeks to assess the risk management impact on the CT performance. Thereby; the decision process for targeting suspect containers is integrated into the functioning of the Container Terminal Management System (CTRMS). CTRMS consists of classified MAS that represent the actors of the CT The CTRMS is classified into six subsystems that regroup agents with similar goals to form coherent groups. The interfacing subsystems with terrestrial and maritime transport providers ensure the generation of the inbound and outbound flows of containers at the terrestrial and maritime interfaces of the CT. The representation subsystem is composed of agents representing the handling equipment existing in the CT and aims to give a graphical aspect to the CT simulation. Planning subsystem ensures the orchestration of the CT handling equipment based on the establishment of a schedule of handling tasks. The supervision subsystem analyze the container description issued from external systems in order to target suspect ones. In addition, it controls the decision of the planning subsystems to check the respect of segregation rules in the storage area. The learning subsystem capitalizes information about containers previously inspected by the custom officers to be used for the extraction of new rules.

To conclude this paper we have describe a framework for intelligent framework for Marine System in respect to CAS the architecture proposed is shown in Figs. 1 and 2. The establishment of a risk management process in a complex system is shown in Fig. 3. Detailed description is given in Sect. 3. To this end, we propose a structured approach to deal with the complexity using multi agent system. In a nutshell, this field of study examines contemporary themes in the seaport risk management complex system and can be implemented in various sea and land ports. It investigates seaport components in research and practice. Multi agent system (MAS) is paradigm well suited to complex problems.

References

1. Barjis, J.: The importance of business process modeling in software systems design. Sci. Comput. Program. **71**(1), 73–87 (2008)
2. Cariou, P., Mejia, M.Q., Wolff, F.C.: Evidence on target factors used for port state control inspections. Marine Policy **33**(5), 847–859 (2009)
3. Chatterjee, A.: An overview of security issues involving marine containers and ports. In: Proceedings of the 2003 Transportation Research Board annual meeting (CD-ROM). Citeseer (2003)
4. Darbra, R.M., Casal, J.: Historical analysis of accidents in seaports. Saf. Sci. **42**(2), 85–98 (2004)
5. Davenport, T.H., Innovation, P.: Reengineering Work Through Information Technology. Harvard Business School Press, Boston (1993)
6. Deen, S.M., Jayousi, R.: A preference processing model for cooperative agents. J. Intell. Inf. Syst. **26**(2), 115–147 (2006). https://doi.org/10.1007/s10844-006-8436-1
7. Ellis, J.: Analysis of accidents and incidents occurring during transport of packaged dangerous goods by sea. Saf. Sci. **49**(8–9), 1231–1237 (2011)
8. Fabiano, B., Currò, F., Reverberi, A.P., Pastorino, R.: Port safety and the container revolution: a statistical study on human factor and occupational accidents over the long period. Saf. Sci. **48**(8), 980–990 (2010)
9. Jayousi, R., Bali, Y.: Querying and semantic mapping approaches for semantic resolution in e-learning system. In: Aggarwal, A., Badra, M., Massacci, F. (eds.) NTMS 2008, 2nd International Conference on New Technologies, Mobility and Security, 5–7 November 2008, Tangier, Morocco, pp. 1–5. IEEE (2008). https://doi.org/10.1109/NTMS.2008.ECP.83
10. Lee, T.W., Park, N.K., Lee, D.W.: Design of simulation system for port resources availability in logistics supply chain. In: Proceedings of the Korean Institute of Navigation and Port Research Conference, pp. 53–61. Korean Institute of Navigation and Port Research (2002)
11. Marhavilas, P.K., Koulouriotis, D., Gemeni, V.: Risk analysis and assessment methodologies in the work sites: on a review, classification and comparative study of the scientific literature of the period 2000–2009. J. Loss Prev. Process Ind. **24**(5), 477–523 (2011)
12. McFarlane, D., Sarma, S., Chirn, J.L., Wong, C., Ashton, K.: Auto ID systems and intelligent manufacturing control. Eng. Appl. Artif. Intell. **16**(4), 365–376 (2003)
13. Meyer, G.G., Främling, K., Holmström, J.: Intelligent products: a survey. Comput. Ind. **60**(3), 137–148 (2009). https://doi.org/10.1016/j.compind.2008.12.005

14. Milazzo, M.F., Ancione, G., Lisi, R., Vianello, C., Maschio, G.: Risk management of terrorist attacks in the transport of hazardous materials using dynamic geoevents. J. Loss Prev. Process Ind. **22**(5), 625–633 (2009)
15. Rattrout, A., Amarneh, R., Badir, H., Fissoune, R.: Improvement of decision making of intelligent products using multi-agent system. In: INTIS 2014, p. 104 (2014)
16. Rattrout, A., Hijawi, H.: Complex adaptive WSNs for polluted environment monitoring. In: ICIT 2015 (2015)
17. Rattrout, A., Yasin, A., Abu-Zant, M., Yasin, M., Dwaikat, M.: Clustering algorithm for AODV routing protocol based on artificial bee colony in MANET. In: Proceedings of the 2nd International Conference on Future Networks and Distributed Systems, pp. 1–9 (2018)
18. Rigas, F., Sklavounos, S.: Risk and consequence analyses of hazardous chemicals in marshalling yards and warehouses at Ikonio/Piraeus harbour, Greece. J. Loss Prev. Process Ind. **15**(6), 531–544 (2002)
19. Ronza, A., Félez, S., Darbra, R., Carol, S., Vílchez, J., Casal, J.: Predicting the frequency of accidents in port areas by developing event trees from historical analysis. J. Loss Prev. Process Ind. **16**(6), 551–560 (2003)
20. Yan, N., Liu, G., Xi, Z.: A multi-agent system for container terminal management. In: 2008 7th World Congress on Intelligent Control and Automation, pp. 6247–6252. IEEE (2008)

Data Item Quality for Biobanks

Vladimir A. Shekhovtsov[ID] and Johann Eder[(✉)][ID]

Alpen-Adria-Universität Klagenfurt, Universitätsstraße 65-67,
9020 Klagenfurt, Austria
{volodymyr.shekhovtsov,johann.eder}@aau.at

Abstract. Biobanks collect and store items of biological material and
provide these resources for medical research together with data associated with these items. In this paper, we contribute to the fundamentals
necessary for establishing data quality management for biobanks. We
analyse the properties of biobanks which are most important for an adequate data quality management system. We provide a comprehensive
description of the concept of quality for biobank data. For this, we state
that the quality of the biobank data can be categorized into data item
quality and metadata quality and provide the detailed treatment of common data item quality characteristics, in particular, completeness, accuracy, reliability, consistency, timeliness, precision, and provenance. These
definitions of data item quality characteristics are a necessary basis for
data quality representation and management. The precise definition of
these data quality characteristics also required as a necessary basis for
integrating data items derived from different sources which is frequently
needed for larger medical studies.

1   Introduction

Biobanks are essential infrastructures for biomedical research. They collect biological material like tissue, blood, etc. and conserve and store this material
with the intention to provide this material for medical studies. In recent years,
biobanks were very successful in the establishment of quality management for
biological specimens such that they can provide material possessing established
quality characteristics [3,10]. Biobanks also provide data associated with the
samples, collecting the data from various sources with an ever increasing demand
and supply of data. We argue that it is necessary to also establish a *quality management system for biobank data*. The reasons for that are as follows:

- Low reproducibility of medical studies has potential dangerous consequences
  for any developments based on these studies such as developing new drugs or
  medical procedures and is at least a waste of resources.

This work has been supported by the Austrian Bundesministerium für Bildung, Wis-
senschaft und Forschung within the project BBMRI.AT (GZ 10.470/0010-V/3c/2018).

A. Hameurlain and A. M. Tjoa (Eds.): TLDKS L, LNCS 12930, pp. 77–115, 2021.
https://doi.org/10.1007/978-3-662-64553-6_5

– To achieve high reproducibility, medical research needs both biomaterial and the associated data of high or at least known quality; the reason is that the reproducibility depends on the reliability of conclusions based on research results, which, in turn, highly depends on the quality of data available for studies.
– Achieving high quality of data is not possible without establishing well defined standard operating procedures for data quality management, and increasing personnel awareness of the importance of such procedures, and the data quality itself.
– In particular, such procedures are necessary because, without them, the resulting quality will only depend on the abilities and the personal circumstances of the persons responsible for data collection.

Establishing such a data quality management system requires that both data quality and how it is measured are properly defined.

Although the recognition of the need for data management in biobanks grows in acceptance [35], the treatment of data quality in the biobank domain is seen as insufficient by the biobank administrators. Precise characterisations of data quality for biobanking are not yet investigated in depth.

The data quality management infrastructure for biobanks has not yet been established in an adequate way, because it lacks the proper background and systematic elaborations of the foundations for data quality for biobanks. In this paper, we contribute to filling this research need and provide the necessary background for establishing such an infrastructure. This paper is based on a set of concepts first presented in [41]. The major contribution of this paper is to provide precise definitions of data quality characteristics and data quality measurements.

This problem is also further reaching. Considerable effort has been spent to establish quality management policies for the treatment and storage of samples (see WP3 of bbmri.at#1 and #2). All these efforts lead to the documentation of the quality of the samples which is of course represented as data associated with the samples.

Searching and collecting relevant cases for medical studies is a laborious and costly process [12]. Therefore, the search infrastructure biobanks and biobank federations offer for researchers should also include documentation of the quality of the data associated with the samples [13,14]. Quality oriented search, however, requires a set of clearly defined quality characteristics which can be used in query conditions.

The rest of the paper is structured as follows. Section 2 presents necessary background information. In Sect. 3, we describe the state of the art in the research on data quality in general. Section 4 provides the conceptualizations and definitions for the data quality in biobanks. In Sect. 5 we precisely define data quality characteristics for the biobank data, and propose the corresponding metrics. In Sect. 6 we draw conclusions and sketch how the introduced definitions can help to establish a data quality management system for biobanks.

2 Background

2.1 Data Quality

We follow [2] in our treatment of data quality as a crucial element of the data management infrastructure. The working definition of data quality accepted for this paper is as follows: Data quality serves as a measure, how well the data represents facts of the real world, and represents the degree to which the data meets the expectations of data consumers based on their intended use of the data. High-quality data meets consumer expectations and represents the real world to a greater degree than low-quality data.

The data quality values can be also interpreted as metadata: as they contain the information describing the data, such as precision, validity period etc.

Assessing the data quality [39] requires understanding the expectations for its use and determining the degree to which the data meets these expectations. Assessment requires understanding of the concepts represented by the data, the processes that created the data, the systems through which the data is created, and the known and potential uses of the data.

2.2 Data Quality Management Systems

In defining the possible components of a quality management system to be established in organizations in the medical domain we follow [36] which defines such system as consisting of:

- An agreed-upon definition of data quality;
- A set of agreed-upon data quality requirements obtained as a result of requirement acquisition;
- A whole set of capable processes for quality assessment, testing and control supplemented with appropriate tools;
- The implementation of common activities for project and personal management.

2.3 Data in Biobanks

Biobanks are important resources for biomedical research, for gaining insights into diseases and for developing therapies and drugs. Biobanks [11,23] are collections of biological material (*samples*) such as tissue, body liquids, cell cultures, etc. accompanied with *data* [35,40]. To provide a definition for such data, we start with following [1] in defining the data in general as "a set of collected facts" of the real world. Based on that, we state that *the biobank data represents a set of collected facts of the real-world samples stored in the biobank*. Such data contains facts about patients and drugs, collected from humans or by means of equipment (in labs, via X-Ray, with microscopes).

We can distinguish the following data-related concepts defined for the biobanks:

1. *Data subjects* are the general categories to which the data belongs i.e. which can be used to classify the data. Such subjects are specified explicitly while establishing the biobank and are used to guide data collection and search. Examples of data subjects can be the *patient data* and the *drugs data*;
2. *Data sources* define possible ways of acquiring data to the biobank. Such sources can belong to different categories (laboratory equipment, X-Ray, human sources);
3. *Data formats* define possible ways of storing and representing the data. Examples of data formats for the biobanks can be textual data, image or video data (X-Ray images, microscopic images etc.)

2.4 Properties of Biobanks Important for Data Quality Management

2.4.1 Categories of Biobanks

We distinguish two main categories of biobanks which can affect data quality treatment: population-based biobanks and clinical biobanks. The position of the specific biobank with respect to these categories affects the treatment of data quality in such a biobank.

1. In the population-based biobanks the data is collected based on the stated criteria from a representative sample of the relevant population.
2. In the clinical biobanks the data is collected from clinics when it is available, often without stated criteria.

A common observation related to these categories is: *it is easier to control the data quality in population-based biobanks*. The reasons for this observation are as follows:

1. The source is known in advance and can be investigated in detail prior to data collection.
2. The collection criteria are stated explicitly beforehand and, as a rule, remain unchanged over the course of the biobank lifecycle.
3. Data quality requirements are defined beforehand and the data acquisition procedures can be designed and established in way to ascertain that these quality criteria are actually achieved.

On the other hand in clinical biobanks the data collection is frequently a byproduct of patient treatment and only data required for this treatment is actually collected. Collections derived from medical study also frequently have high data quality but the data quality is focused on the requirements of a specif study (or groups of studies).

Further in this paper, we show how the category of biobank influences the set of selected quality characteristics for its data.

2.4.2   Data Sources for Biobanks

To introduce the data quality for biobanks, it is necessary to know which kinds of data is stored there. We start from classifying the data by its source, distinguishing three categories of data sources:

1. the biobank itself;
2. the external sources;
3. the external parties processing the biobank data.

These categories are explained in detail below. Further in this paper, we will establish the set of quality characteristics for the sources belonging to every category.

Data Coming from the Biobank Itself. Such biobank-originated data can belong to the following categories:

1. *The sample handling data.* Such data is continuously produced and recorded as a part of the regular biobank sample handling processes. Examples of such data are ischemia time and storage temperature.
2. *The data explicitly produced for collections.* An example is the data from the cohort studies where the biomaterial is explicitly collected together with the data to be provided for research.

Data Coming from External Sources. The externally produced data can belong to the following categories:

1. *The data produced by scientific studies.* Such data is added to the biobank because the materials used in these studies are stored there. An example of such data is the donor data collected by means of disease-based collection from ontology, when the typical approach is to have such data collected in the clinical departments and not by the biobanking unit.
2. *The data produced by routine healthcare.* The most important category of this data is the data from Electronic Health Records (EHRs). An example is the information from the patient health record such as height, weight, or anamnesis: the whole health record is typically not produced by the biobank but is used there.
3. *The data from linked collections.* Such data comes from any data collections containing the data produced elsewhere but used in the biobank (e.g. by linking to such collections from the materials in the biobank).

Feedback Data. Such data is supplied back to the biobank in a feedback loop, when the external parties paper back their findings based on the data originated from the biobank itself. For example, the data produced from the biobank materials can be given to scientific studies which analyze this data, produce some new data as a result of such analysis, and this data is reported back to the biobank.

3   State of the Art

In this section, we review the state of the art within the scope of this paper.

We deal with the body of work investigating the data quality in our domain. Within the medical domain, we start from briefly reviewing the body of work dealing with data quality in medical research in general, without going into specifics. After that, we address the body of work devoted to two specific topics within this domain which are the closest to our scope: (1) the quality of electronic health records, and (2) the quality of data in biobanks. By treating the research on electronic health records separately, we emphasize their importance as data sources for biobanks.

3.1   Data Quality in Medical Information Systems

3.1.1   Data Quality in Medical Research Not Specific to Biobanks

This body of research addresses data quality in the healthcare domain in general, without addressing the biobanks explicitly. In [36], different aspects of the data quality in clinical studies are addressed in detail, such as available data sources, and data quality characteristics which are specific to the domain. Based on that, the authors present a practical framework for dealing with data quality in the healthcare domain, which can be further specialized to cover specific healthcare subdomains. In this paper, we largely follow this route in establishing the data quality framework for the biobank domain.

A significant body of research deals with specific issues within clinical data quality domain, some of them being applicable to biobanking. We can distinguish

1. The research on the *quality of data sources* in general. In [19], the authors described their attempts at conducting source data verification audit procedures on the health care datasets to monitor data quality, claiming as a result that such audits could identify data quality and integrity problems within such datasets.
2. The research on *data quality management procedures*. The procedures for strategic management of data quality in healthcare are proposed in [25]. In [20], the authors established an approach to defining a set of quality monitoring procedures to ensure data integrity based on investigation of state of the art in data quality management in clinical trials. [17] deals with the aspects of monitoring data quality in controlled trials.
3. The research on the *specific data quality characteristics*. Among such characteristics are data reliability [18,21], usefulness [9,52], and completeness [30,31].

3.1.2   Quality of Electronic Health Records (EHRs)

The research on the quality of electronic health records is within the scope of this paper, as such records can serve as data sources for biobanks.

General Research Sources. We start from the body of work covering wide sets of aspects of EHR quality with various degree of comprehension.

The survey of the state of the art in this domain is provided in [51], it covers both EHR quality characteristics and the methods for data quality assessment. [8,16] provide further reviews of EHR quality characteristics and quality assessment methods. [22] deals with EHR data quality from the IS perspective.

Dealing with Specific EHR Quality Characteristics. The following table groups the research papers by high-level quality characteristic they deal with. The list of characteristics is based on the list presented in [51]: completeness, correctness, concordance, plausibility, and currency. As stated in that paper:

1. the treatment of completeness also includes availability (unavailable records cannot be considered incomplete);
2. the treatment of correctness also includes dealing with accuracy and validity (incorrect records are invalid and not accurate);
3. the treatment of concordance also includes reliability and consistency;
4. plausibility can be also referred to as believability and truthworthness;
5. currency is a synonym for timeliness.

Characteristic	How it is defined
Completeness	[31] explicitly by means of a conceptual model together with the approach for developing completeness hypotheses; [50] as non-random completeness of EHRs; [49,51] as completeness of EHRs intended for reuse;
Correctness	[51] as the ability to provide true information; [47] as compliance to clinical guidelines; [38] as the ability to preserve medical information;
Concordance	[6] as reliability of quality measures based on electronic records; [29] as consistency in the coding system used in EHR; [38] as internal and external consistency: internal consistency preserves identical identifiers for a new EHR version for the same patients, external consistency preserves such identifiers across patients
Plausibility	[15] as a probability that a diagnosis information in the EHR corresponds to the true disease
Currency	[4] as timeliness serving as a basis for the correctness and fairness of conclusions drawn from EHRs

3.1.3 Data Quality in Biobanks

The research on data quality in the biobank domain can be traced back to [11] which emphasized the importance of provenance as an important factor for data quality as it documents the process of data integration; it did not address further quality characteristics. The research on provenance in medical domain with partial applicability to biobanks was continued in [46].

Some further works address different aspects of data quality in biobanks without establishing a comprehensive framework. In particular, in [5], achieving high data quality for biobanks is considered equally important to reaching high sample quality: "there is a need to define sample and data quality, define best practice for biobanks and set up a scheme to confirm that biobanks are following best practice guidelines and achieving high quality". Unfortunately, this work does not show how to reach such goal in practice. In [28], the treatment of data quality is limited to the quality of the cancer registry data to be linked to biobanks, though the importance of data quality in general is also understood: "in biobank-based studies, data quality concerns are present in each phase of the study - from the donation of the biological specimen, to the long-term storage of study files".

Finally, in [41], the authors introduced the detailed set of data quality characteristics together with the idea of treating the biobank as a data broker, and the data quality in the biobank as the metadata quality. Still, no attempts at establishing the practical framework for the biobank people aiming at dealing with data quality in a consistent manner was made at that time. An additional complication comes from the anonymization [7] of medical data which might be necessary to be able to share the data. Anonymization typically requires to generalize the data [45] which causes, for an example a reduction of precision.

To sum it up, despite the extensive research on the topic of data quality in the medical domain in general, the subject of data quality for biobanks was not covered extensively enough. In our paper, we aim at addressing this void.

4   Data Quality Definition for Biobanks

As mentioned above, one of the components for the quality management system is an agreed-upon definition of the data quality. In achieving this, it is necessary to agree on a set of data quality characteristics. This is the goal of this section.

4.1   Biobanks as Data Brokers

We start from concentrating on the quality of the data produced by the external sources, as opposed to the data which is produced by and for the biobanks. The biobanks frequently serve as *data brokers*, using and publishing such externally produced data. They provide matching (mediating) services and descriptions in case of incomplete data.

From the quality point of view, the difference is that for the data produced by and for the biobank, the biobank has to ensure the data quality *itself*, whereas for the data produced elsewhere, the biobank has to ensure a correct *description* of the data quality. The metaphor for such description process can be the one of a shop selling consumer goods (e.g. a camera): the shop is not responsible for the quality of the camera, but responsible for the right (full, unambiguous etc.) description of the camera and dealing with its quality problems (e.g. measuring, accepting complaints etc.).

For this case, the standard definition of data quality as a fitness for use for a particular purpose is not applicable. The reason is that the biobank primarily collects the samples for future research studies, and not for a particular purpose, so the data associated with these samples is not known a priori. As a result, it is not possible to know in advance both the scientific purposes for which the collected samples can be used, and the quality requirements for not-yet-formed data in these studies. This also leads to problems while dealing with the informed consent, as providing such consent in advance is not always possible.

In fact, the question related to the fitness for use must be answered differently for biobanks. To get to such an answer, we need to know the intended usage for the biobank data. This data, as we stated before, includes both sample descriptions and the descriptions of the data associated with the samples. Such usage is twofold:

1. The data is used for deciding whether samples and data can be used for an intended study.
2. The data is used to aid in the search for suitable samples with certain characteristics. Such search can be performed against the set of the quality descriptions of both the samples and the data associated with the samples.

Here we can make a connection to the library management domain, as in a library, the search also looks into the descriptions of books, and not into the books themselves. As it is known that the library is mostly responsible for the quality of the descriptions of the books, and not in quality of the books themselves (as only the former is under their control, helping to find the books), we assume that this is also true for the biobanks. Based on all that, the data quality can be interpreted as *fitness for search*.

4.2 Introducing Quality Characteristics for Biobank Data

Quality characteristics help in achieving proper understanding of data quality, reflecting different aspects of this quality. Such characteristics can be supplemented with measures, such measures allow for quantification of the degree of achievement the particular quality. The characteristics can be combined to obtain the integrated (overall) data quality.

Quality metrics allow for quantifying quality characteristics, they represent a measurement process resulting in the specific values.

4.2.1 Categorizing Quality Characteristics

Based on the distinction between the assessment levels introduced above, it is possible to state that the biobank data quality characteristics fall into two main categories.

1. *Data item quality characteristics* assess the quality of the data item level:
   (a) the quality of data coming from external sources;
   (b) the quality of data produced by and for the biobanks;
   (c) the quality of feedback data;
2. *Metadata quality characteristics* assess the quality of the metadata level:
   (a) the quality of data item characteristics treated as metadata;

(b) the quality of non-quality-related metadata.

In this paper, we deal with data item quality characteristics, metadata quality characteristics will be a subject of the subsequent paper. Prior to proceeding with the definition of the data item characteristics, however, we introduce the concept of biobank data quality, which includes both data item and metadata quality.

4.2.2 Conceptualizing Biobank Data Quality

Based on the previous discussion, we propose the conceptualization of the biobank data quality presented on Fig. 1.

Within this conceptual schema, the *Biobank Data Quality* concept is comprised of a set of *Biobank Data Quality Characteristics* each of which can be either *Data Item*, or *Metadata Quality Characteristic*. Every such Characteristic is assessed by means of a set of corresponding *Metrics*. Applying a *Data Item Quality Metric* to measure quality of a specific *Biobank Data Item* results in a *Data Item Quality Measurement* which produces a quality assessment value forming a *Quality Metadata Element* (as we treat such values as metadata) connected back to the corresponding Biobank Data Item.

It is also possible to apply a *Metadata Quality Metric* to measure quality of a Quality Metadata Element (second-level quality assessment), this results in a *Metadata Quality Measurement* which also produces a quality assessment value forming another Quality Metadata Element, this time connected back to the original Quality Metadata Element.

Both metadata element-data item and metadata element-metadata element connections are handled by generalizing Biobank Data Item and Quality Metadata Element up to a *Biobank Data Element* (as we treat metadata as a kind of data itself), and allowing Biobank Data Elements to connect to each other.

For the sake of simplicity, we assume that biobank data items are not categorized further as external, internal, or feedback data items.

4.2.3 Intrinsic and Relative Data Item Quality Characteristics and Metrics

While presenting the quality characteristics by defining their metrics, following [48], we distinguish between relative and intrinsic metrics and treat them separately. Our goal in separating these two groups of metrics is that we are going to establish the separate framework for collecting measurement data based on intrinsic metrics, such framework will be introduced in the next section.

Intrinsic quality metrics are defined as follows: *the quality assessment based on the intrinsic quality metric must only rely on the existing biobank data.* The following rules apply:

1. Using intrinsic metrics to perform measurements must not rely on human opinions, such metrics *are not subjective*. For example, the metrics for understandability or perceived reliability are relative by definition as they are subjective. Consequently, it is only possible for relative metrics to be based on

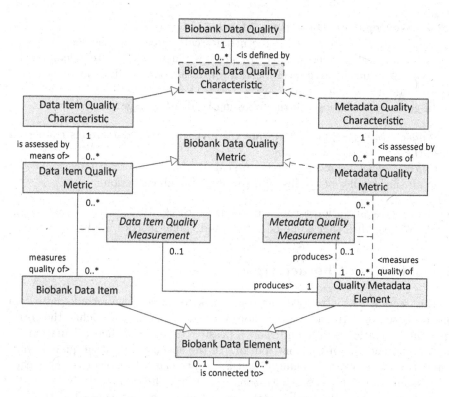

Fig. 1. Conceptualization of the biobank data quality

goals to be fulfilled by the biobank data or the biobank itself, as these goals are set by humans based on subjective judgements.

2. intrinsic metrics must not rely on external e.g. standard data (such as the data used by benchmarks or other external reference data), as all measurement values have to be calculated based on the existing data within the biobank.
3. intrinsic metrics must not refer to the data schema, only data values have to be considered. For example, the metrics for data schema completeness are relative, as well as the metrics comparing the data schema with some external e.g. standard schemas.

When the metrics which can be defined for a quality characteristic are exclusively intrinsic or relative, it is possible to introduce *intrinsic* or *relative quality characteristics*. For example, data understandability is a relative quality characteristic, as it is not possible to define intrinsic understandability metrics. We also distinguish *mixed quality characteristics* which have a mixed set of metrics (some being intrinsic, and some relative). For such characteristics, we list only intrinsic metrics within our framework, and introduce the rest later.

4.2.4 Aggregating Data Item Metrics

Instead of dealing directly with the metrics calculated over the separate data items, it is possible to use *aggregated metrics*. Such metrics are calculated by aggregating values of item-level metrics for data items of different granularity, such as samples or collections.

In general, we distinguish the following kinds of aggregate metrics:

1. *sample-level aggregate metrics* are calculated over all data items connected to the specific biobank sample;
2. *collection-level aggregate metrics* are calculated over all data items connected to all the samples belonging to a specific biobank collection.

An exact aggregation formula depends on the kind of metrics being aggregated, we deal with that while describing individual metrics.

5 Data Item Characteristics

In this section, we introduce our framework for assessing biobank data item quality based on intrinsic metrics. For every characteristic, we define the corresponding intrinsic metrics, and show a possible example of their calculation.

We concentrate on the contributions of the specific data item quality characteristics to the overall quality of data. Quality characteristics can affect each other, some positively (boosting), some negatively (hindering). In Sect. 5.8, we present an example illustrating such interdependent quality characteristics.

5.1 Data Item Completeness

This characteristic [26,33] reflects the need in collecting all required data. Sufficient completeness (contributing to high quality) usually means that all such data is present. For medical data, insufficient completeness is detected when some data attributes are missing because they were either not recorded due to unavailability, or simply not transmitted. Measuring completeness can take into account the relative importance of attributes, as some attributes can be more problematic to miss. Most of the data completeness metrics are intrinsic.

5.1.1 Declaring Data Item Attributes

We define data item completeness for the biobank domain based on the fragment of the conceptual schema shown on Fig. 2.

This schema distinguishes between *declared* and instantiated (*possessed*) data attributes. There are two ways to declare data attributes for a collection:

1. Following [14], the content of a collection can be characterized by a set of the ontological concepts belonging to the reference ontology, e.g. such concepts can define information on the data collection method, the time aspect of the data collection, the item or property observed etc. The ontological concept can

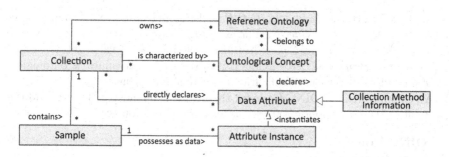

Fig. 2. A fragment of the collection conceptual schema related to data completeness

refer to the declarations of presence of the data item attributes. For example, if the content of the collection is characterized by a concept defining the data collection method, it entails the presence of the *Collection Method* attribute for the data items belonging to this collection. We further refer to such case as the case of the ontological concept *declaring* the data item attribute.

2. The collection description can directly declare specific data item attributes (e.g. when no ontological concepts are available at all). For example, it again can declare the Collection Method attribute and contain the set of names of the allowed data collection methods. Such attributes are declared exclusively for a specific collection.

5.1.2   Declaring Data Item Attributes with LOINCs

In this section, we show an example of declaring data item attributes when a particular specialization of the reference ontology is used to describe collection contents. In this, we again follow [14] to specialize such an ontology by means of the *LOINC* standard (Logical Observation Identifiers Names and Codes (https:// loinc.org)) and the ontological concepts - by means of *LOINC Concepts*.

LOINC provides a code system for identifying laboratory and clinical observations and test results (such as vital signs, level of blood hemoglobin etc.) The LOINC Code (a unique identifier defined by a LOINC standard) can be substituted for the *Coded Information* structured to contain several *LOINC Parts* (https://loinc.org/download/loinc-users-guide). Out of these parts, we distinguish the following subset as being able to declare data item attributes (in the list below, we reuse their descriptions provided in [14]):

1. *Component*: the name of the measured component or analyte (e.g., glucose, propranolol); for diagnostic tests, it can refer to the disease information e.g. its ICD code;
2. *Property*: the property which was observed as a result of a test or observation (e.g., substance concentration, mass, volume);
3. *System*: the type of a system or sample (e.g., urine, serum, or the patient as a whole);
4. *Method*: the method of the measurement (e.g., radioimmunoassay, immune blot); this part is optional.

A LOINC Part may hold a *LOINC Part Value* belonging to a domain represented by *LOINC Part Value Set*. For example a *Property* part can contain the value of "*CCnc (Catalytic Concentration)*" belonging to a value set of "*Enzymatic Activity Properties*". We follow [14] in treating such values as ontological concepts and value sets as specifications of a Reference Ontology.

An example of the LOINC Code for the primary diagnosis ICD code and the fragment of the corresponding Coded Information is as follows:

– **LOINC Code:** *86255-7*
– **Coded Information:**
  • **Component:** *Primary diagnosis ICD code*
  • **Property:** *Type (Known Type)*
  • **System:** *Patient (Patient as a whole).*

It is possible to distinguish two ways of using LOINC concepts to declare data item attributes which correspond to two ways of attaching LOINCs to the collection data defined in [14]:

1. the specific LOINC Code can describe the content of the specific Collection declaring it as matching the corresponding Coded Information as a whole; in this case the data item attribute can be declared based on the values of any subset of the present LOINC parts.
2. the specific LOINC Part Value together with the information to which LOINC Part it belongs, can describe the content of the specific Collection; in this case the data item attribute can be declared based on the value of this LOINC part.

Declaring the data item attribute based on the value of the LOINC part can be done

1. *directly* - with the attribute definition having one-to-one correspondence to the part value; for example, the value of "Primary diagnosis ICD code" for a Component LOINC part can directly declare the data item attribute named *PrimaryDiagnosisICDCode* to contain the values of such codes.
2. *indirectly* - with the attribute definition indirectly derived from one or several part values.

5.1.3 Defining Completeness Based on Declared Data Attributes
The samples belonging to a collection have to instantiate (possess) the declared data attributes. *Failing to do so means that the data is incomplete.*

1. If the data item attributes are declared by an ontological concept (e.g. a LOINC concept), we can assume that such attributes have to be instantiated for the samples belonging to all collections characterized by such concept. For example, if the concept connected to the collections A and C declares the *Collection Method* attribute, we can assume that such concept declares that the data for the samples belonging to both A and C instantiates this

attribute, and will be incomplete if it is missing. If there are more than one ontological concept connected to a collection, the resulting set of declared attributes for such collection can be defined as follows:

$$A_c = \bigcup_{k \in K_c} A_k,$$

where K_c is a set of all ontological concepts connected to a collection c, A_k is a set of all attributes declared by a specific ontological concept k.

2. The same is true for the data attributes declared as a part of the collection description. For example, if such description declares the *Collection Method* data item attribute, we can again assume that it declares that the sample data will be incomplete if this attribute is not present.

For the sake of simplicity, we assume that a sample instantiates a single value for a specific attribute, i.e. we do not consider multi-valued attributes.

5.1.4   Sample Completeness

We define sample completeness as *a degree of attribute instantiation for a sample.*

Sample Completeness Based on a Simple Presence Function. The sample completeness in the simplest "presence-based" case (reflecting only the fact of the presence of the filled data item representing the attribute value) can be calculated as a ratio of a number of instantiated attributes to a total number of attributes declared for a collection to which the sample belongs:

$$SCP_s = \frac{n(A_{C(s)}) - \sum\limits_{a \in A_{C(s)}} f_s(a)}{n(A_{C(s)})}$$

where $A_{C(s)}$ is a set of all attributes defined for a collection $C(s)$ to which the sample s belongs, $n(A_{C(s)})$ is a cardinality of this set, $f_s(a)$ is a negative presence function over attribute $a \in A_{C(s)}$ which can be either 0 if a is present in s, or 1 otherwise. For example, if there are 12 missing attributes out of 100 in s, $SCP_s = (100 - 12)/100 = 0.88$.

Sample Completeness Based on a Reverse Completeness Function. If it makes sense to reflect not only the fact of presence, but also the degree of completeness for the data attribute value, the negative presence function can be replaced with the reverse completeness function which is actually a normalized distance in completeness space between $a \in A_{C(s)}$ and the area defined by a full completeness.

$$SCC_s = \frac{n(A_{C(s)}) - \sum\limits_{a \in A_{C(s)}} h_s(a) / \max\limits_{a \in A_{C(s)}} (h_s(a))}{n(A_{C(s)})}$$

where $h_s(a)$ is a distance between the instance of the attribute $a \in A_{C(s)}$ defined for the sample s, and the area defined by full completeness (i.e. where all the points are fully complete attribute values). If it is only possible to estimate the completeness for some attribute values as belonging to a given interval, this interval estimation can be substituted for all these values: for example, if 25 out of 100 attribute values are within 10% of the maximum recorded distance outside the full completeness interval (normalized $h_s(a) = 0.1$), 20 are near 100% of that distance (normalized $h_s(a) = 1$) and the rest are fully complete ($h_s(a) = 0$), the completeness is calculated as follows: $SCC_s = (100 - (25 \cdot 0.1 + 20 \cdot 1))/100 = 0.73$.

5.1.5   Collection Completeness

Such completeness is calculated over the whole collection. We distinguish the following types of collection completeness:

1. *Sample-based collection completeness* is calculated based on the completeness values for the samples belonging to a collection;
2. *Attribute-level collection completeness* is calculated over the values for a specific attribute defined for all samples in a collection;
3. *Subset-based collection completeness* is calculated over the values for a specific subset of attributes defined for all samples in a collection;
4. *Full collection completeness* is calculated over the values for all attributes defined for all samples in a collection.

For brevity, we do not consider here the completeness calculated for a subset of samples in a collection (e.g. belonging to a specific subcollection or meeting specific criteria). Also, we restrict our definitions of collection-level completeness metrics by using only a simple presence function, though it is also possible to define such metrics using a reverse completeness function.

Sample-Based Collection Completeness. Such completeness is calculated by applying the aggregate function (e.g. average, minimum/maximum, or median) to a set of completeness values calculated for all the samples belonging to a given collection. For example, *sample-average-based collection completeness* can be calculated as follows:

$$CCP_c^{samp} = \frac{\sum\limits_{s \in S_c} SCP_{sc}}{n(S_c)},$$

where S_c is a set of all samples in a collection c, SCP_{sc} is a value of a sample completeness calculated for a sample s in c.

Attribute-Scoped Collection Completeness. We define such completeness as *a degree of value presence for a specific attribute in a specific collection*. It can be calculated as follows:

$$CCP_{ac}^{attr} = \frac{n(S_c) - \sum\limits_{s \in S_c} f_a(s)}{n(S_c)}$$

where S_c is a set of all samples belonging to a collection c, $n(S_c)$ is a cardinality of this set, $f_a(s)$ is a negative attribute presence function over a sample

$s \in S_c$ which can be either 0 if the value for the data attribute a is present in s, or 1 otherwise. For example, if a is missing in 12 out of 100 samples in c, $CCP_{ac}^{attr} = (100 - 12)/100 = 0.88$.

Subset-Scoped Collection Completeness. To support queries, where somebody is searching for a collection with a set of specific attributes present, it is possible to define subset-scoped collection completeness as *a degree of value presence for a specific subset of attributes in a specific collection.*

We define several possible approaches for calculating subset-based completeness:

1. it can be calculated by applying the aggregate function (e.g. average, minimum/maximum, or median) to a set of attribute-level completeness values defined for all attributes in a subset. For example, *average-based collection completeness* can be calculated as follows:

$$CCP_c^{avgS}(T) = \frac{\sum\limits_{a \in T} CCP_{ac}^{attr}}{n(T)}, T \subseteq A_c$$

where $T \subseteq A_c$ is a subset of attributes defined for a collection c, A_c is a set of all attributes defined for c.

2. it can be the result of applying the reverse attribute subset presence function $f_T^{all}(s)$ defined over all attributes in a subset. Such function indicates if at least one of the attributes in a subset is missing for a given sample in a collection. With such function, the collection completeness can be calculated as follows:

$$CCP_c^{allS}(T) = \frac{n(S_c) - \sum\limits_{s \in S_c} f_T^{all}(s)}{n(S_c)}, T \subseteq A_c$$

where S_c is a set of all samples belonging to a collection c, $n(S_c)$ is a cardinality of this set, $f_T^{all}(s)$ is a reverse attribute subset presence function over a sample s which returns 0 if all attributes from T are present in s, and 1 if at least one of them is missing. This way, $CCP_c^{allS}(T)$ indicates *the degree of full subset completeness* i.e. the ratio of the number of samples possessing all attributes in a set to the total number of samples. For example, if at least one of the attributes from T is missing in 12 out of 100 samples in c (and the rest is complete with respect to T), $CCP_c^{allS}(T) = (100 - 12)/100 = 0.88$.

3. It can be the result of applying the full subset emptiness function $f_T^{any}(s)$ which returns 0 if at least one of the attributes belonging to T is present in s, or 1 if all attributes are missing. This way, the collection completeness calculated as

$$CCP_c^{anyS}(T) = \frac{n(S_c) - \sum\limits_{s \in S_c} f_T^{any}(s)}{n(S_c)}, T \subseteq A_c$$

indicates *the degree of partial subset completeness*, i.e. the ratio of the number of samples possessing at least one of the attributes in a set to the total number of samples. For example, if all attributes from T are missing in 12 out of 100 samples in c (and the rest is at least partially complete with respect to T), $CCP_c^{anyS}(T) = (100 - 12)/100 = 0.88$. Another useful metric in this case is *the degree of complete subset emptiness* calculated as $1 - CCP_c^{anyS}(T)$ i.e. the ratio of the number of samples with all subset attributes missing to the total number of samples.

4. It can be also the result of applying the relative attribute presence function $f_T^{ratio}(s)$ which returns the ratio of the number of non-empty attributes belonging to T to the total number of attributes in T for a given sample s. This way, the collection completeness can be calculated as

$$CCP_c^{ratioS}(T) = \frac{n(S_c) - \sum\limits_{s \in S_c} f_T^{ratio}(s)}{n(S_c)}, T \subseteq A_c$$

and takes into account the number of non-empty attributes for every sample.

Full Collection Completeness. Such completeness is defined as *a degree of value presence for all attributes in a specific collection*. It is in fact a specific case of the previous characteristic when $T = A_c$ and can be calculated using the formulas presented above. In particular

1. *average-based full collection completeness* is calculated as follows:

$$CCP_c^{avg} = \frac{\sum\limits_{a \in A_c} CCP_{ac}^{attr}}{n(A_c)}$$

where A_c is a set of all attributes declared for a collection c.

2. *the degree of full sample completeness* is calculated as follows

$$CCP_c^{all} = \frac{n(S_c) - \sum\limits_{s \in S_c} f_c^{all}(s)}{n(S_c)}$$

where S_c is a set of all samples belonging to a collection c, $n(S_c)$ is a cardinality of this set, $f_c^{all}(s)$ is a reverse all attribute presence function over a sample $s \in S_c$ which can be either 0 if all attributes belonging to A_c are present in s, or 1 if at least one of them is missing. For example, if at least one of the attributes from c is missing in 12 out of 100 samples in c, and the rest of the samples possess the complete set of attribute values, $CCP_c = (100 - 12)/100 = 0.88$.

5.2   Data Accuracy and Validity

Data accuracy reflects the need to represent the real world truthfully [37]. Low accuracy means that the data is vague and not precise enough, or plainly incorrect (not corresponding to reality). High quality data is accurate.

Data accuracy is a mixed quality characteristic as some accuracy metrics depend on the human judgement or the external sample data to be used in comparisons. Here we only list metrics which can be expressed as intrinsic metrics.

In this section, we distinguish between

1. *syntactic accuracy metrics* which reflects the ability of the data itself to correspond to the domain constraints,
2. *indirect accuracy metrics* which reflect the accuracy of the context elements associated with the data, in particular, diagnostic methods used for collecting the data.

We follow [27] in treating *data validity* as a quality characteristic which is based on a specific subset of indirect data accuracy metrics so it is partially synonymous with accuracy. We discuss this together with discussing the relevant accuracy metrics.

5.2.1 Syntactic Accuracy Metrics

Such metrics measure the degree of correspondence between the data items and the constraints related to a given domain. An example of the data item violating the domain constraint can be the negative value recorded for the patient age. The list of such metrics is as follows.

1. *Syntactic accuracy based on a simple threshold function.* We follow [44] in establishing the metric for the syntactic accuracy which is a "closeness of a value to the elements of the corresponding definition domain" [44]. Such syntactic accuracy in the simplest "threshold-based" case (reflecting only the fact of the violations of the domain constraints), is calculated as the following aggregated metric for the data items belonging to the same domain:

$$Syna_d = \frac{n(I_d) - \sum\limits_{i \in I_d} f_d(i)}{n(I_d)}, d \in D$$

where D is a set of all domains for the data items under consideration (e.g. in a collection), $n(I_d)$ is a cardinality of the set I_d which includes all data items belonging to the domain D (or their fixed-size subset e.g. 100 items), $f_d(i)$ is a negative threshold function over data item $i \in I_d$ related to its domain $d \in D$ which can be either 0 if i corresponds to constraints of d, or 1 otherwise. An example of the domain constraint served well by such metric, is the non-negativity constraint for the age domain, in the case when negative age values are unacceptable. For example, if the constraints of the domain d are violated in 12 cases out of 100, $Syna_d = (100 - 12)/100 = 0.88$.

2. *Syntactic accuracy based on reverse closeness function.* If it makes sense to reflect not only the fact, but also the degree of domain constraint violations for the data item, the negative threshold function can be replaced with the reverse closeness function which is actually a normalized distance between $i \in I_d$ and the area defined by the constraints of its domain d.

$$Syna_d = \frac{n(I_d) - \sum\limits_{i \in I_d} h_d(i) / \max\limits_{i \in I_d}(h_d(i))}{n(I_d)}, d \in D$$

where $h_d(i)$ is a distance between the data item $i \in I_d$ and the area defined by the constraints of its domain d. An example of the domain constraint served well by such metric, is the specific acceptance interval for the results of a given measurement, when the values inside the interval are completely acceptable, and for the values outside the interval, the distance still influences the accuracy. If it is only possible to estimate the distance for some items as belonging to a given interval, this interval estimation can be substituted for all these items: for example, if 15 out of 100 measurements are within 10% of the maximum recorded distance outside the acceptance interval (normalized $h_d(i) = 0.1$), 10 are near 100% of that distance (normalized $h_d(i) = 1$) and the rest are inside the interval ($h_d(i) = 0$), the accuracy is calculated as follows: $Syna_d = (100 - (15 \cdot 0.1 + 10 \cdot 1))/100 = 0.885$.

Aggregating Syntactic Accuracy Metrics. To calculate the aggregated syntactic accuracy metric for a collection, it is necessary to apply the aggregated function (e.g. a minimum, mean or median) to all $Syna_d, d \in D_C$, where D_C is a set of all domains belonging to a collection C. E.g. the mean syntactic accuracy is calculated as follows:

$$Syna_C = \frac{\sum\limits_{d \in D_C} Syna_d}{n(D_C)}$$

Out of the above functions, the result of applying the minimum function indicated the worst accuracy recorded for a collection, i.e. the "bottleneck" in terms of syntactic accuracy.

Using LOINCs to Define Accuracy Metrics. The LOINC Code can be a substitute for the information including the set of allowed values for a concept e.g. a set of allowed ICD codes. If LOINC information declares the data item attribute, such set can define a domain for this attribute. In this case the violation of the domain constraint can take the form of the data item containing the value which does not belong to such a set.

5.2.2   Method-Based Indirect Accuracy Metrics

If we include context into consideration, it becomes possible to establish indirect accuracy metrics. For such metrics, accuracy has to be measured indirectly, i.e. the metric has to be based not on the measurement of the data item itself, but on the inherent, accumed, or measured quality of an element (or elements) of its context, and the quality of the data item has to be derived from that context quality.

One of such context elements is *the method used for collecting or producing the data*, so, one of the approaches to reduce the set of metrics to intrinsic metrics, is to base the accuracy assessment on the accuracy of such method. For example, the data for the immunity status against measles will be more accurate if collected by means of data analysis rather than by asking patients to recall if they suffered from measles as children. Similarly, the data for the tentative diagnosis will be less accurate if this diagnosis is made through a fast test rather than by means of a thorough test because the fast tests are designed to have very little false negatives, while precise tests should minimize the sum of false positives and false negatives.

Based on the above discussion, we propose a set of item-level intrinsic metrics to assess data accuracy. Such metrics are atomic, as they are defined on data items which are not decomposed further, so they do not aggregate other metrics applied to the results of such decomposition.

Accuracy Metrics for Diagnostic Methods. We start with measuring accuracy characteristics of the producing diagnostic methods. These metrics are applicable for data items which are the result of applying the diagnostic method where the estimated number of the expected outcomes for the given method is known. An example of such data item is the type of the disease (diagnosis).

We follow [32] by defining the following five data accuracy metrics based on the accuracy of the diagnostic method:

1. *Method sensitivity.* Such metric indicates "the ability of a test to detect the disease when it is truly present" [32] and can be calculated as follows:

$$Sens = \frac{TP}{TP + FN},$$

   where TP is the number of true positive cases, FN is the number of false negative cases.

2. *Method specificity.* Such metric indicates "the probability of a test to exclude the disease status in patients who do not have the disease" [32] and can be calculated as follows:

$$Spec = \frac{TN}{TN + FP},$$

   where TN is a number of true negative cases, FP is the number of false positive cases.

3. *Method positive predictive value (PPV).* Such metric indicates "the probability that a patient has the disease given that the test results are positive" [32] and can be calculated as follows:

$$PPV = \frac{TP}{TP + FP},$$

4. *Method negative predictive value (NPV).* Such metric indicates "the probability that a patient does not have the disease given that the test results are indeed negative" [32] and can be calculated as follows:

$$NPV = \frac{TN}{TN + FN},$$

5. *Method likelihood ratio (LR)*. Such metric "indicates the ratio of the probability of the test result among patients who truly had the disease to the probability of the same test among patients who do not have the disease, <...­ its magnitude> informs about the certainty of a positive diagnosis" [32]. It can be calculated as follows:

$$LR = \frac{Sens}{100 - Spec},$$

Method Validity as a Specific Kind of Method Accuracy. We follow [27] in stating that some of the above metrics also characterize validity of the diagnostic methods, and, consequently, the validity of the data. According to [27] "validity is described in terms of sensitivity <...> and specificity", so we propose the following definition: *method validity is a specific kind of method accuracy based on the subset of accuracy metrics including method sensitivity and method specificity.*

As it is clear that the way of connecting method validity metrics to the data items will be the same as for other accuracy metrics, we will not separately deal with method and data validity further in this paper.

Connecting Method Accuracy Metrics to Samples and Collections. After defining method accuracy metrics the next step is to connect these metrics to the biobank data structures. We introduce several ways to declare diagnostic methods for a collection:

1. As we discussed before, collections are characterized by ontological concepts (e.g. represented by LOINCs), where each ontological concept corresponds to a structure which could include the information on the data collection methods. As a result, the data collection methods connected to a ontological concept are declared for all collections characterized by this concept.
2. It is also possible to attach ontological concepts to the specific attributes declared for a collection, in this case the data collection methods connected to an ontological concept are declared for all data attributes characterized by this concept;
3. The collection description can directly declare specific data collection methods. The scope of such methods is limited to this collection.
4. The information on the specific data collection methods can be also directly attached to the specific attributes described for a collection. The scope of such methods is limited to these attributes.

The samples belonging to a collection have to instantiate the declared diagnostic method information for their attributes as follows:

1. If the data collection methods are declared in the information connected to an ontological concept, and this concept characterizes some collections, we can assume that the information on such methods have to be instantiated for the values of at least some of the attributes possessed by the samples belonging to all collections characterized by such concept. If there are more than one concept connected to a collection, the resulting set of diagnostic methods for such collection can be defined as follows:

$$M_c = \bigcup_{k \in K_c} M_k,$$

where K_c is a set of all ontological concepts connected to a collection c, M_k is a set of all data collection methods declared by a specific concept k.

2. If the ontological concept characterizes specific attributes declared for a collection, the information on its data collection methods has to be instantiated for the values of exactly these attributes of the samples belonging to this collection.

3. If the diagnostic methods are declared as a part of the collection description, these methods have to be instantiated for at least some of the values of the attributes of the samples belonging to this collection.

4. If the diagnostic methods characterize specific attributes declared as a part of the collection description, these methods have to be instantiated for the values of exactly these attributes of the samples belonging to this collection.

Based on the above, we distinguish the following sample-level aggregate method accuracy metrics:

1. *the average degree of method sensitivity for a specific sample*: such metric is applicable when the values for different data attributes possessed by the given sample are collected using different methods. It is calculated over all attribute values belonging to a sample for which the method information is available:

$$Sens_s^{avg} = \frac{\sum\limits_{a \in A_{C(s)}} Sens(m_a(s))}{n(A_{C(s)})},$$

where $A_{C(s)}$ - a set of all data attributes declared for the collection $C(s)$ containing the sample s, for which the collection method information is available, $Sens(m_a(s))$ is a sensitivity of the diagnostic method $m_a(s)$ used for collecting the value for the attribute a belonging to a sample s.

2. *the median method sensitivity for a specific sample*: it reflects the method sensitivity value which is most likely to be encountered in practice for a given sample;

3. *the minimum of method sensitivity for a specific sample*: it reflects the "worst case" i.e. the most problematic sensitivity recorded for the sample;

4. *the mean, the median, and the minimum of method specificity, PPV, NPV, and method likelihood ratio for a specific sample*;

We distinguish the following collection-level aggregate accuracy metrics (for brevity, we limit ourselves here only by metrics based on averages, with one exception):

1. *the sample-based average degree of method sensitivity for a collection*: it is calculated as an average of the sensitivity values for all samples in a collection:

$$Sens_c^{avgS} = \frac{\sum\limits_{s \in S_c} Sens_s^{avg}}{n(S_c)},$$

   where S_c is a set of all samples in c.

2. *the attribute-scoped average degree of method sensitivity for a collection*, such metric makes sense, if the values for the same attribute for different samples were collected by different methods:

$$Sens_a^{avgattr} = \frac{\sum\limits_{s \in S_c} Sens(m_a(s))}{n(S_c)}, a \in A_c$$

3. *the attribute-based average degree of method sensitivity for a collection*: it is calculated as an average of the attribute-scoped sensitivity values for all attributed declared for a collection, for which the method information is available:

$$Sens_c^{avgA} = \frac{\sum\limits_{a \in A_c} Sens_a^{avgattr}}{n(A_c)},$$

4. *the absolute minimum for method sensitivity for a collection*: such metric defines the narrowest "bottleneck" for a method sensitivity which exists in the whole collection:

$$Sens_c^{\min S} = \min_{s \in S_c} Sens_s^{\min},$$

where $Sens_s^{\min}$ is a minimum method sensitivity for a sample s.

5.2.3   Blending Method Accuracy Metrics with Expert Knowledge

It is possible to blend together accuracy metrics based on expert judgement collected beforehand. The resulting metric is defined on the ratio scale between 0 an 1 (1 indicating the highest possible accuracy for the method), the values are calculated based on expert judgements over the alternative methods in terms of the quality criteria such as the method accuracy metrics defined above. Again, the judgement value for the specific method is propagated as data accuracy metric values to all the data items connected to this method.

In converting expert judgements into ratio values, it is possible to rely on such methods as Analytic Hierarchy Process (AHP) [42] or Analytic Network Process [43]. Following [24], we outline the sequence of steps to be followed to calculate ratio values for the compared methods by means of AHP based on quality criteria defined as atomic accuracy metrics.

1. Define the main objective of the comparison, this time selecting the most accurate data collection method;
2. Define the set of characteristics of the methods, which can be used for comparison. As mentioned above, the method accuracy metrics i.e. its sensitivity, specificity etc. can serve as such characteristics.
3. Pick the alternatives, i.e. the methods under consideration. Together with the characteristics (accuracy metrics) they build the complete AHP hierarchy.
4. Rank the methods in terms of characteristics (accuracy metrics), e.g. calculate the sensitivity for all the methods and rank them based on its values.
5. Allow the experts to rank the relative importance of the characteristics for the case at hand. Such ranking is performed by means of pairwise comparison in terms of the objective. Such comparison can e.g. involve answering the questions like "which characteristic of the pair is more important for the accuracy of the method?".
6. Blend the ranking values obtained in steps 4 and 5 by means of AHP to obtain the final ranking.

The results of the final method ranking are used as metric values for the data items connected to the corresponding methods.

This technique can be also used in situations, when ranking the methods in terms of the values of their accuracy metrics is not possible e.g. as the information about the number of their false positives etc. is not available, or these methods are not diagnostics by nature (e.g. they just collect the data from end users). In this case such ranking is performed by means of pairwise comparison of the alternatives in terms of characteristics. Such comparison can e.g. involve answering the questions like "which method of the pair possesses higher sensitivity?".

5.3   Data Reliability

This characteristic characterize the underlying measured concept. For the medical data related to a concept (such as e.g. the depressive mood), reliability can be defined as a degree to which a question triggering the collection of this data, represents a reliable measure of that concept. Another possible definition derives reliability of the data from the reliability of its source i.e. to its provenance. For example, the reliability of the data about the cause of deaths is much higher if it was provided by a trained pathologist than if it was attributed to a general practitioner. Low reliability and validity mean that the data cannot be trusted, so its quality is low.

5.3.1   Method-Based Indirect Reliability Metrics

We propose to treat data reliability similarly to data accuracy by defining *indirect data reliability metrics* based on the reliability metrics defined for diagnostic methods used for data collection. The rest of this section is structured following the description of indirect data accuracy metrics in Sect. 5.2.2.

Reliability Metrics for Diagnostic Methods. According to [27], the reliability of the diagnostic method "refers to <its> capacity <...> to give the same result on repeated application."

The most widely used reliability metrics for diagnostic methods are test-retest coefficients and split-half measures [27].

Connecting Method and Source Reliability Metrics to Samples and Collections. The approach for connecting method and source reliability metrics to samples and collections is the same as the approach described in Sect. 5.2.2 for method accuracy metrics, so we only show here some resulting sample-level and collection-level data reliability metrics.

An example of sample-level aggregate reliability metric is *the average degree of method repeatability for a specific sample*: such metric is applicable when the values for different data attributes possessed by the given sample are collected using different methods. It is calculated over all attribute values belonging to a sample for which the method information is available:

$$Rept_s^{avg} = \frac{\sum\limits_{a \in A_{C(s)}} Rept(m_a(s))}{n(A_{C(s)})},$$

where $A_{C(s)}$ - a set of all data attributes declared for the collection $C(s)$ containing the sample s, for which the collection method information is available, $Sens(m_a(s))$ is a repeatability of the diagnostic method $m_a(s)$ used for collecting the value for the attribute a belonging to a sample s.

Some examples of collection-level aggregate reliability metrics are:

1. *the sample-based average degree of method repeatability for a collection*: it is calculated as an average of the repeatability values for all samples in a collection:

$$Rept_c^{avgS} = \frac{\sum\limits_{s \in S_c} Rept_s^{avg}}{n(S_c)},$$

   where S_c is a set of all samples in c.

2. *the attribute-scoped average degree of method repeatability for a collection*, such metric makes sense, if the values for the same attribute for different samples were collected by different methods:

$$Rept_a^{avgattr} = \frac{\sum\limits_{s \in S_c} Rept(m_a(s))}{n(S_c)}, a \in A_c$$

3. *the absolute minimum for method repeatability for a collection*: such metric defines the narrowest "bottleneck" for a method repeatability which exists in the whole collection:

$$Rept_c^{\min S} = \min_{s \in S_c} Rept_s^{\min},$$

where $Rept_s^{\min}$ is a minimum method repeatability for a sample s.

5.3.2 Blending Method Reliability Metrics with Expert Knowledge

As for accuracy, it is also possible to blend together accuracy metrics based on expert judgement. The process described in Sect. 5.2.3 can be applied to method reliability metrics. As a result, the sequence of steps described in that section, will look like this:

1. Define the main objective of the comparison, this time selecting the most reliable data collection method;
2. Define the set of characteristics of the methods, which can be used for comparison. The method reliability metrics can serve as such characteristics.
3. Pick the alternatives, i.e. the methods under consideration. Together with the characteristics (reliability metrics) they build the complete AHP hierarchy.
4. Rank the methods in terms of characteristics (reliability metrics);
5. Allow the experts to rank the relative importance of the characteristics for the case at hand. Such ranking is performed by means of pairwise comparison in terms of the objective. Such comparison can e.g. involve answering the questions like "which characteristic of the pair is more important for the reliability of the method?".
6. Blend the ranking values obtained in steps 4 and 5 by means of AHP to obtain the final ranking.

The results of the final method ranking are used as metric values for the data items connected to the corresponding methods.

As for accuracy, this technique can be also used in situations, when ranking the methods in terms of the values of their reliability metrics is not possible e.g. as the necessary information is not available, or these methods are not diagnostics by nature. In this case such ranking is performed by means of pairwise comparison of the alternatives in terms of characteristics. Such comparison can e.g. involve answering the questions like "which method of the pair is more reliable?".

5.4 Data Consistency

This characteristic reflects the need for non-conflicting data. For the medical data sufficient consistency (contributing to high quality) means that there is no contradiction in the data as the real-world states reflected by data items are not in conflict.

In this paper, data consistency is measured by sample-level and collection-level metrics as a reverse degree of variability with respect to the data collection method within a sample or a collection. This means, for example, that *the collection data is more consistent if it was collected by smaller number of methods*. We start with sample-level metrics, and then continue with collection-level metrics.

5.4.1 Sample-Level Consistency

On the sample level the possible metric can be a *sample consistency degree based on the total number of methods* calculated as a reverse ratio of the total number of methods used to collect attribute values for a sample to the total number of methods used to collect attribute values for its collection.

$$Cons_s^{totS} = 1 - \frac{n(M_s)}{n(M_{c(s)})},$$

where M_s is a set of methods used to collect the data for the sample s, $M_c(s)$ is a set of methods used to collect the data for the collection $c(s)$ to which the sample s belongs. Note that here we count the number of methods really used to collect the data as opposed to the methods just declared e.g. by ontological concepts or in a collection description. Here we assume that only one method can be used to collect a specific attribute value for a sample.

It is also possible to take into account the number of values collected by the methods. An example of such metric can be a *sample consistency degree based on the most frequently applied method* calculated as the the ratio of the number of values collected by the most frequently applied method to the total number of collected values:

$$Cons_s^{freqS} = \frac{\max_{m \in M_s} n(V_{ms})}{n(V_s)},$$

where V_{ms} - the set of attribute values collected by the method m for the sample s, V_s - the total set of attribute values collected for the sample s. Higher values of such metric mean that higher percentage of values was collected by a single method.

5.4.2   Collection-Level Consistency

On the collection level, we distinguish the following consistency metrics:

1. *the sample-based average consistency degree for a collection*: it is calculated as an average of the consistency values for all samples in a collection:

$$Cons_c^{avgtotS} = \frac{\sum\limits_{s \in S_c} Cons_s^{totS}}{n(S_c)}$$

   or

$$Cons_c^{avgfreqS} = \frac{\sum\limits_{s \in S_c} Cons_s^{freqS}}{n(S_c)},$$

   where S_c is a set of all samples in c.

2. *the attribute-scoped consistency degree for a collection based on number of methods* which is calculated based on the number of methods used to collect all values for a specific attribute in a collection:

$$Cons_a^{totA} = 1 - \frac{n(M_a)}{\max\limits_{a' \in A_c} n(M_{a'})}, a \in A_c,$$

where M_a is a set of methods used to collect all data values for the attribute a, A_c is a set of all attributes declared for a collection c.

3. *the attribute-scoped consistency degree for a collection based on most frequently applied method* which is calculated based on the ratio of the number of values for a specific attribute in a collection collected by most frequently used method to the total number of values collected to this attribute:

$$Cons_a^{freqA} = \frac{\max\limits_{m \in M_a} n(V_{ma})}{n(V_a)}, a \in A_c$$

where M_a is a set of methods used to collect all data values for the attribute a, V_{ma} is a set of values collected by the method m for the attribute a for the whole collection, V_a is a set of values collected for the attribute a for the whole collection.

4. *the attribute-based average consistency degree for a collection*: it is calculated as an average of the attribute-scoped consistency values for all attributes declared for a collection:

$$Cons_c^{avgtotA} = \frac{\sum\limits_{a \in A_c} Cons_a^{totA}}{n(A_c)},$$

or

$$Cons_c^{avgfreqA} = \frac{\sum\limits_{a \in A_c} Cons_a^{freqA}}{n(A_c)},$$

5. *the minimum data consistency for a collection*: such metric defines the narrowest "bottleneck" for a data consistency which exists in the whole collection:

$$Cons_c^{\min totS} = \min\limits_{s \in S_c} Cons_s^{totS}$$

or

$$Cons_c^{\min freqS} = \min\limits_{s \in S_c} Cons_s^{freqS}$$

5.5   Data Timeliness

This characteristic reflects the need for the data to represent actual real-world states. It characterizes the collection process and can be defined as a length of time between the change in a real-world state and the time when the data reflects that change. Close to timeliness is **Currency** which is related to the data itself: as a result of timely collection, the data is kept current i.e. not obsolete. Good quality data is current and is collected in timely manner (shortly after observation). An example for timeliness could be the importance of measuring BMI body weight recently as old measurements do not represent the truth anymore.

The sample is accompanied by a time when it was collected, whereas every its attribute value is accompanied by a creation time. It makes sense to look at the distance between these two types of values. We define data timeliness as *the reverse distance of time between creating the sample and adding the data.* An example of low timeliness is the case when collecting BMI in done in two years after collecting the blood sample: such situations are usually unacceptable.

5.5.1   Sample-Level Timeliness

An example of sample-level aggregate timeliness metric is *the average timeliness for a specific sample.*

$$
Timel_s^{avg} = \frac{\sum\limits_{a\in A_{C(s)}} 1 - d_s(a)/\max\limits_{a\in A_{C(s)}} d_s(a)}{n(A_{C(s)})},
$$

where $A_{C(s)}$ - a set of all data attributes declared for the collection $C(s)$ containing the sample s, for which the creation time information is available, $d_s(a)$ is a distance between the creation time of the sample s and the creation time of its attribute a.

Minimum timeliness for a specific sample could indicate a sample-level timeliness bottleneck:

$$
Timel_s^{min} = \min_{a\in A_{C(s)}} 1 - d_s(a)/\max_{a\in A_{C(s)}} d_s(a)
$$

5.5.2   Collection-Level Timeliness

Some examples of collection-level aggregate timeliness metrics are:

1. *sample-based average collection-level timeliness*: it is calculated as an average of the timeliness values for all samples in a collection:

$$
Timel_c^{avgS} = \frac{\sum\limits_{s\in S_c} Timel_s^{avg}}{n(S_c)},
$$

where S_c is a set of all samples in c.

2. *attribute-scoped average collection-level timeliness*:

$$
Timel_a^{avgattr} = \frac{\sum\limits_{s\in S_c} 1 - d_s(a)/\max\limits_{s\in S_c} d_s(a)}{n(S_c)}, a \in A_c
$$

3. *the absolute minimum for sample timeliness for a collection*: such metric defines the narrowest "bottleneck" for a sample timeliness which exists in the whole collection:

$$
Timel_c^{min\,S} = \min_{s\in S_c} Timel_s^{min},
$$

where $Timel_s^{min}$ is a minimum timeliness for a sample s calculated by means of minimum function instead of average.

5.6   Data Precision

This characteristic can be interpreted as the number of significant digits for number variables or as the rule resolution of the categories – for categorical variables. For example, having just three categories for blood pressure is obviously not very precise and contributes to quality negatively. Other examples of precision are the scale for the ischemia time (which can be specified in hours, minutes, or seconds) or the precision of blood pressure measurement equipment.

We define data precision as *the degree of category resolution for the values of categorical data attributes and the number of significant digits - for the values of numeric data attributes.* An example of the degree of category resolution can be the number of used categories.

5.6.1   Sample-Level Precision

On the sample level we can define the following possible metrics:

1. an *average precision degree based on the total number of categories* calculated as an average of the number of categories used for the attribute values for a sample.

$$Prec_s^{avgtotK} = \frac{\sum\limits_{a \in A_c^K(s)} n(K_a) / \max\limits_{a \in A_c^K(s)} n(K_a)}{n(A_{c(s)}^K)},$$

where K_a is a set of data categories used for the attribute a, $A_c^K(s)$ is a set of categorical attributes defined for the collection $c(s)$ to which the sample s belongs. Note that here we count the number of categories used for the data as opposed to the methods just declared e.g. by ontological concepts or in a collection description. Here we assume that only one category can be used for a specific attribute value for a sample.

2. an *average precision degree based on the number of significant digits* calculated as an average of the number of significant digits used for the attribute values for a sample.

$$Prec_s^{avgsigd} = \frac{\sum\limits_{a \in A_c^G(s)} g_s(a) / \max\limits_{a \in A_c^G(s)} g_s(a)}{n(A_{c(s)}^G)},$$

where $g_s(a)$ is a function returning the number of significant digits for a value for the attribute a for the sample s, $A_c^G(s)$ is a set of attributes with significant digits defined for the collection $c(s)$ to which the sample s belongs.

3. a *minimum precision degree based on the total number of categories* which allows to find the precision bottleneck for a sample

$$Prec_s^{mintotK} = \min\limits_{a \in A_c^K(s)} (n(K_a) / \max\limits_{a \in A_c^K(s)} n(K_a))$$

4. a *minimum precision degree based on the number of significant digits* calculated as

$$Prec_s^{minsigd} = \min\limits_{a \in A_c^G(s)} (g_s(a) / \max\limits_{a \in A_c^G(s)} g_s(a))$$

5.6.2   Collection-Level Precision

On the collection level, we distinguish the following precision metrics:

1. *the sample-based average precision degree for a collection*: it is calculated as an average of the precision values for all samples in a collection:

$$Prec_c^{avgtotKS} = \frac{\sum\limits_{s \in S_c} Prec_s^{avgtotK}}{n(S_c)}$$

or

$$Prec_c^{avgsigdS} = \frac{\sum\limits_{s \in S_c} Prec_s^{avgsigd}}{n(S_c)}$$

where S_c is a set of all samples in c.

2. *the attribute-scoped precision degree for a collection* which is calculated based on the number of categories or the number of significant digits used for a specific attribute in a collection:

$$Prec_a^{avgtotKA} = \frac{\sum\limits_{s \in S_c} n(K_a)/\max\limits_{s \in S_c} n(K_a)}{n(S_c)}, a \in A_c^K$$

$$Prec_a^{avgsigdA} = \frac{\sum\limits_{s \in S_c} g_s(a)/\max\limits_{s \in S_c} g_s(a)}{n(S_c)}, a \in A_c^G$$

It is also possible to use minimum functions instead of averages.

3. *the attribute-based average precision degree for a collection*: it is calculated as an average of the attribute-scoped precision values for all attributes declared for a collection:

$$Prec_c^{avgtotKA} = \frac{\sum\limits_{a \in A_c^K} Prec_a^{avgtotKA}}{n(A_c^K)},$$

or

$$Prec_c^{avgsigdA} = \frac{\sum\limits_{a \in A_c^G} Prec_a^{avgsigdA}}{n(A_c^G)},$$

4. *the minimum data precision for a collection*: such metric defines the narrowest "bottleneck" for a data precision which exists in the whole collection:

$$Prec_c^{\min KS} = \min\limits_{s \in S_c} Prec_s^{\min totK}$$

or

$$Prec_c^{\min sigdS} = \min\limits_{s \in S_c} Cons_s^{\min sigd}$$

5.7   Data Provenance

This characteristic [46] is the degree of linking between the data sources and collection methods on the one side, and the data values on the other side. For low-provenance data (contributing to low quality), it is not possible or difficult to understand where it comes from or how it was collected.

We propose the following definition of this characteristic: *the data provenance is a completeness with respect to collection method or source*. Based on that definition, we distinguish *data source provenance* and *collection method provenance*. In this section, we will only consider collection method provenance, the approach for collecting data source provenance is similar.

We define collection method provenance as a degree of completeness with respect to the information about the data collection method. It answers the following question: *is it known how the data was collected, which method was used?* It is important to note that for defining such characteristic, we are only interested in the information about the methods connected to the samples, not in the declarations of the methods e.g. by means of ontological concepts.

5.7.1   Sample-Level Provenance

We define sample-level provenance as *a degree of method information instantiation for a sample*. It can be calculated as a ratio of a number of instantiated collection methods for the attributes to a total number of attributes declared for a collection to which the sample belongs:

$$Prov_s^{samp} = \frac{n(A_{C(s)}) - \sum\limits_{a \in A_{C(s)}} f_s^m(a)}{n(A_{C(s)})}$$

where $A_{C(s)}$ is a set of all attributes defined for a collection $C(s)$ to which the sample s belongs, $n(A_{C(s)})$ is a cardinality of this set, $f_s^m(a)$ is a negative method presence function over attribute $a \in A_{C(s)}$ which can be either 0 if the information on the collection method is present for a in s, or 1 otherwise. For example, if the information on collection method is missing for 12 attributes out of 100 in s, $SCP_s = (100 - 12)/100 = 0.88$.

5.7.2   Collection-Level Provenance

Such provenance is calculated over the whole collection. We distinguish the following types of collection-level provenance:

1. *Sample-based collection-level provenance* is calculated based on the provenance values for the samples belonging to a collection;
2. *Attribute-scoped collection-level provenance* is calculated over the values for a specific attribute defined for all samples in a collection;
3. *Subset-based collection-level provenance* is calculated over the values for a specific subset of attributes defined for all samples in a collection;

4. *Full collection-level provenance* is calculated over the values for all attributes defined for all samples in a collection.

Sample-Based Collection-Level Provenance. Such completeness is calculated by applying the aggregate function (e.g. average, minimum/maximum, or median) to a set of provenance values calculated for all the samples belonging to a given collection. For example, *sample-average-based collection-level provenance* can be calculated as follows:

$$Prov_c^{avgS} = \frac{\sum\limits_{s \in S_c} Prov_{sc}^{samp}}{n(S_c)},$$

where S_c is a set of all samples in a collection c, $Prov_{sc}^{samp}$ is a value of a sample provenance calculated for a sample s in c.

Attribute-Scoped Collection-Level Provenance. We define such completeness as *a degree of value presence for a specific attribute in a specific collection*. It can be calculated as follows:

$$Prov_{ac}^{attr} = \frac{n(S_c) - \sum\limits_{s \in S_c} f_a^m(s)}{n(S_c)},$$

where S_c is a set of all samples belonging to a collection c, $n(S_c)$ is a cardinality of this set, $f_a(s)$ is a negative method presence function over a sample $s \in S_c$ which can be either 0 if the information about the collection method is present for the data attribute a in s, or 1 otherwise. For example, if the information about collection method for the attribute a is missing in 12 out of 100 samples in c, $Prov_{ac}^{attr} = (100 - 12)/100 = 0.88$.

Subject-Scoped and Full Collection-Level Provenance. It is also possible to collect subset-scoped and full collection-level provenance metrics in a way shown for subset-scoped and full completeness in Sect. 5.1.5, we will not deal with it here in detail.

5.8   Example of Dealing with Quality Characteristics

To illustrate the possible ways of dealing with interdependent quality characteristics we use a problem of *defining the volume of data*, which should be recorded for specimens for future research.

This problem is closely related to data quality. For example, in digital pathology, the pathologists need a resolution of images which is way beyond what is commonly available in smartphones or cheaper digital cameras: the visual *precision* of such images is very high. The problem is that collecting images of such resolution creates enormous amount of data. As a result, it is necessary to decide, often in real time, which image data to keep and which to recompute or record again from the same material.

This question is quite difficult to answer. To address it, it is necessary to introduce the quality characteristic of *cost-effectiveness*. Based on its definition, we can state that, to arrive to a cost-effective solution, it is necessary to set up a system which can decide, based on the precision requirements and the available resources, for which case and subset of samples which amount of data should be recorded. For example, if the processing is routine, resulting in the large number of similar samples accompanied with the large amount of similar data in the database, the decision could be that it is not necessary to store such data. If, to the contrary, the case is not "normal" i.e. very special, rare – then the images or videos have to be recorded in a much higher resolution.

A solution can be to establish a process for making such discard-or-store decisions. Establishing such decision process is complicated because the decision could affect the data which becomes necessary in a quite distant future (e.g. in 10 years), but which is not necessary right now. And usually it is not possible to know what will be needed in 10 years' time, whereas recording everything right now (achieving the highest possible precision) increases the cost significantly and can be not affordable at all. Following the definition of cost-effectiveness, we can state that the answer to this question depends on the context but is usually related to making the best use of available resources. It can also make use of the forecasts of future needs and requirements.

6   Conclusions

The overarching goal for the management activities in biobanks as a part of the quality management system for samples, processes, and data as proposed in this paper is the reproducibility of scientific studies based on the material and data provided by a biobank. The problem is that, for supporting scientific studies with data, it is necessary to know about the quality of that data. The quality characteristics are inherent for the biobanks data (i.e. the data always has qualities); the question is how to be aware of these qualities. For example, while all the data possess some precision, it is not always possible to know how precise or imprecise the specific data item is. To achieve this, it is important to have the data quality *documented* and *assessed*. Such documentation is important also because it gives the necessary quality-related information prior to combining data from different sources.

For establishing a data quality management system and to provide information about data quality information for researchers searching for material and data for their study, it is of great importance that the data quality properties are precisely defined. In this paper we contributed to these foundations for an adequate data quality system.

The *quality of the data quality documentation* is also important because such documentation helps researchers to decide whether it is possible to use the specific material or the specific data set for some research project. To increase the overall quality, the data quality documentation should use interchangeable reference ontologies. Achieving such interchangeability is an important task for international bodies (such as BBMRI-ERIC [34]). To get there, they have to find the agreement on ontologies they are going to use, if such agreement is found, the members of such bodies can collaborate over institutional boundaries more easily.

References

1. ASQ Quality Glossary. https://asq.org/quality-resources/quality-glossary/d
2. Batini, C., Scannapieco, M.: Data and Information Quality: Dimensions, Principles and Techniques. Springer, Cham (2016). https://doi.org/10.1007/978-3-319-24106-7
3. Betsou, F.: Quality assurance and quality control in biobanking. In: Hainaut, P., Vaught, J., Zatloukal, K., Pasterk, M. (eds.) Biobanking of Human Biospecimens, pp. 23–49. Springer, Cham (2017). https://doi.org/10.1007/978-3-319-55120-3_2
4. Cao, S., Zhang, G., Liu, P., Zhang, X., Neri, F.: Cloud-assisted secure eHealth systems for tamper-proofing EHR via blockchain. Inf. Sci. **485**, 427–440 (2019)
5. Carter, A., Betsou, F.: Quality assurance in cancer biobanking. Biopreserv. Biobank. **9**(2), 157–163 (2011)
6. Chan, K.S., Fowles, J.B., Weiner, J.P.: Electronic health records and the reliability and validity of quality measures: a review of the literature. Med. Care Res. Rev. **67**(5), 503–527 (2010)
7. Ciglic, M., Eder, J., Koncilia, C.: Anonymization of data sets with null values. In: Hameurlain, A., Küng, J., Wagner, R., Decker, H., Lhotska, L., Link, S. (eds.) Transactions on Large-Scale Data- and Knowledge-Centered Systems XXIV. LNCS, vol. 9510, pp. 193–220. Springer, Heidelberg (2016). https://doi.org/10.1007/978-3-662-49214-7_7
8. Cowie, M.R., et al.: Electronic health records to facilitate clinical research. Clin. Res. Cardiol. **106**(1), 1–9 (2017)
9. Dinov, I.D.: Volume and value of big healthcare data. J. Med. Stat. Inf. **4** (2016)
10. Dollé, L., Bekaert, S.: High-quality biobanks: pivotal assets for reproducibility of OMICS-data in biomedical translational research. Proteomics **19**(21–22), 1800485 (2019)
11. Eder, J., Dabringer, C., Schicho, M., Stark, K.: Information systems for federated biobanks. In: Hameurlain, A., Küng, J., Wagner, R. (eds.) Transactions on Large-Scale Data- and Knowledge-Centered Systems I. LNCS, vol. 5740, pp. 156–190. Springer, Heidelberg (2009). https://doi.org/10.1007/978-3-642-03722-1_7
12. Eder, J., Gottweis, H., Zatloukal, K.: It solutions for privacy protection in biobanking. Public Health Genom. **15**(5), 254–262 (2012)
13. Eder, J., Shekhovtsov, V.A.: Data quality for medical data lakelands. In: Dang, T.K., Küng, J., Takizawa, M., Chung, T.M. (eds.) FDSE 2020. LNCS, vol. 12466, pp. 28–43. Springer, Cham (2020). https://doi.org/10.1007/978-3-030-63924-2_2
14. Eder, J., Shekhovtsov, V.A.: Data quality for federated medical data lakes. Int. J. Web Inf. Syst. (2021). Publisher: Emerald Publishing Limited
15. Estiri, H., Vasey, S., Murphy, S.N.: Generative transfer learning for measuring plausibility of EHR diagnosis records. J. Am. Med. Inform. Assoc. **28**, 559–568 (2020)
16. Feder, S.L.: Data quality in electronic health records research: quality domains and assessment methods. West. J. Nurs. Res. **40**(5), 753–766 (2018)

17. Fougerou-Leurent, C., et al.: Impact of a targeted monitoring on data-quality and data-management workload of randomized controlled trials: a prospective comparative study. Br. J. Clin. Pharmacol. **85**(12), 2784–2792 (2019)
18. Götzinger, M., Anzanpour, A., Azimi, I., TaheriNejad, N., Rahmani, A.M.: Enhancing the self-aware early warning score system through fuzzified data reliability assessment. In: Perego, P., Rahmani, A.M., TaheriNejad, N. (eds.) Mobi-Health 2017. LNICST, vol. 247, pp. 3–11. Springer, Cham (2018). https://doi.org/10.1007/978-3-319-98551-0_1
19. Houston, L., Probst, Y., Humphries, A.: Measuring data quality through a source data verification audit in a clinical research setting. Stud. Health Technol. Inform. **214**, 107–13 (2015)
20. Houston, L., Probst, Y., Yu, P., Martin, A.: Exploring data quality management within clinical trials. Appl. Clin. Inform. **9**(01), 072–081 (2018)
21. Huzooree, G., Khedo, K.K., Joonas, N.: Data reliability and quality in body area networks for diabetes monitoring. In: Maheswar, R., Kanagachidambaresan, G.R., Jayaparvathy, R., Thampi, S.M. (eds.) Body Area Network Challenges and Solutions. EICC, pp. 55–86. Springer, Cham (2019). https://doi.org/10.1007/978-3-030-00865-9_4
22. Jetley, G., Zhang, H.: Electronic health records in IS research: quality issues, essential thresholds and remedial actions. Decis. Support Syst. **126**, 113137 (2019)
23. Karimi-Busheri, F., Rasouli-Nia, A.: Integration, networking, and global biobanking in the age of new biology. In: Karimi-Busheri, F. (ed.) Biobanking in the 21st Century. AEMB, vol. 864, pp. 1–9. Springer, Cham (2015). https://doi.org/10.1007/978-3-319-20579-3_1
24. Kaschek, R., Pavlov, R., Shekhovtsov, V.A., Zlatkin, S.: Characterization and tool supported selection of business process modeling methodologies. In: Abramowicz, W., Mayr, H.C. (eds.) Technologies for Business Information Systems, pp. 25–37. Springer, Dordrecht (2007). https://doi.org/10.1007/1-4020-5634-6
25. Kerr, K.A., Norris, T., Stockdale, R.: The strategic management of data quality in healthcare. Health Informatics J. **14**(4), 259–266 (2008)
26. Király, P., Büchler, M.: Measuring completeness as metadata quality metric in Europeana. In: 2018 IEEE International Conference on Big Data (Big Data), pp. 2711–2720. IEEE (2018)
27. Kyriacou, D.N.: Reliability and validity of diagnostic tests. Acad. Emerg. Med. **8**(4), 404–405 (2001)
28. Langseth, H., Luostarinen, T., Bray, F., Dillner, J.: Ensuring quality in studies linking cancer registries and biobanks. Acta Oncol. **49**(3), 368–377 (2010)
29. Lee, D., Jiang, X., Yu, H.: Harmonized representation learning on dynamic EHR graphs. J. Biomed. Inform. **106**, 103426 (2020)
30. Liu, C., Talaei-Khoei, A., Zowghi, D., Daniel, J.: Data completeness in healthcare: a literature survey. Pac. Asia J. Assoc. Inf. Syst. **9**(2) (2017). ISBN 1943-7544
31. Liu, C., Zowghi, D., Talaei-Khoei, A., Daniel, J.: Achieving data completeness in electronic medical records: a conceptual model and hypotheses development. In: Proceedings of the 51st Hawaii International Conference on System Sciences (2018)
32. Mandrekar, J.N.: Simple statistical measures for diagnostic accuracy assessment. J. Thorac. Oncol. **5**(6), 763–764 (2010)
33. Margaritopoulos, M., Margaritopoulos, T., Mavridis, I., Manitsaris, A.: Quantifying and measuring metadata completeness. J. Am. Soc. Inform. Sci. Technol. **63**(4), 724–737 (2012)

34. Mayrhofer, M.T., Holub, P., Wutte, A., Litton, J.E.: BBMRI-ERIC: the novel gateway to biobanks. Bundesgesundheitsblatt-Gesundheitsforschung-Gesundheitsschutz **59**(3), 379–384 (2016)
35. Müller, H., Dagher, G., Loibner, M., Stumptner, C., Kungl, P., Zatloukal, K.: Biobanks for life sciences and personalized medicine: importance of standardization, biosafety, biosecurity, and data management. Curr. Opin. Biotechnol. **65**, 45–51 (2020)
36. Nahm, M.: Data quality in clinical research. In: Richesson, R., Andrews, J. (eds.) Clinical Research Informatics, pp. 175–201. Springer, London (2012). https://doi.org/10.1007/978-1-84882-448-5_10
37. Olson, J.E.: Data Quality: The Accuracy Dimension. Morgan Kaufmann, San Francisco (2003)
38. Pantazos, K., Lauesen, S., Lippert, S.: De-identifying an EHR database-anonymity, correctness and readability of the medical record. In: MIE, pp. 862–866 (2011)
39. Pipino, L.L., Lee, Y.W., Wang, R.Y.: Data quality assessment. Commun. ACM **45**(4), 211–218 (2002)
40. Quinlan, P.R., Gardner, S., Groves, M., Emes, R., Garibaldi, J.: A data-centric strategy for modern biobanking. In: Karimi-Busheri, F. (ed.) Biobanking in the 21st Century. AEMB, vol. 864, pp. 165–169. Springer, Cham (2015). https://doi.org/10.1007/978-3-319-20579-3_13
41. Ranasinghe, S., Pichler, H., Eder, J.: Report on data quality in biobanks: problems, issues, state-of-the-art. arXiv preprint 1812.10423 (2018)
42. Saaty, T.L.: Decision making with the analytic hierarchy process. Int. J. Serv. Sci. **1**(1), 83–98 (2008)
43. Saaty, T.L., Vargas, L.G.: Decision Making with the Analytic Network Process, vol. 282. Springer, Boston (2006). https://doi.org/10.1007/978-1-4614-7279-7
44. Salati, M., et al.: Task-independent metrics to assess the data quality of medical registries using the European Society of Thoracic Surgeons (ESTS) Database. Eur. J. Cardiothorac. Surg. **40**(1), 91–98 (2011)
45. Stark, K., Eder, J., Zatloukal, K.: Priority-based k-anonymity accomplished by weighted generalisation structures. In: Tjoa, A.M., Trujillo, J. (eds.) DaWaK 2006. LNCS, vol. 4081, pp. 394–404. Springer, Heidelberg (2006). https://doi.org/10.1007/11823728_38
46. Stark, K., Koncilia, C., Schulte, J., Schikuta, E., Eder, J.: Incorporating data provenance in a medical CSCW system. In: Bringas, P.G., Hameurlain, A., Quirchmayr, G. (eds.) DEXA 2010. LNCS, vol. 6261, pp. 315–322. Springer, Heidelberg (2010). https://doi.org/10.1007/978-3-642-15364-8_26
47. Staroselsky, M., et al.: Improving electronic health record (EHR) accuracy and increasing compliance with health maintenance clinical guidelines through patient access and input. Int. J. Med. Informatics **75**(10–11), 693–700 (2006)
48. Stvilia, B., Gasser, L., Twidale, M.B., Shreeves, S.L., Cole, T.W.: Metadata quality for federated collections. In: Proceedings of the Ninth International Conference on Information Quality (ICIQ 2004), pp. 111–125 (2004)
49. Weiskopf, N.G., Hripcsak, G., Swaminathan, S., Weng, C.: Defining and measuring completeness of electronic health records for secondary use. J. Biomed. Inform. **46**(5), 830–836 (2013)
50. Weiskopf, N.G., Rusanov, A., Weng, C.: Sick patients have more data: the non-random completeness of electronic health records. In: AMIA Annual Symposium Proceedings, vol. 2013, p. 1472. American Medical Informatics Association (2013)

51. Weiskopf, N.G., Weng, C.: Methods and dimensions of electronic health record data quality assessment: enabling reuse for clinical research. J. Am. Med. Inform. Assoc. **20**(1), 144–151 (2013)
52. Zúñiga, F., Blatter, C., Wicki, R., Simon, M.: National quality indicators in Swiss nursing homes: questionnaire survey on data reliability and users' view on the usefulness. Z. Gerontol. Geriatr. **52**(8), 730–736 (2019)

Author Index

Printed in the United States
by Baker & Taylor Publisher Services